MW01325098

SOURCE
The Prentice Hall
ENGINEERING SOURCE

Excel XP/2002 for Engineers

David C. Kuncicky
Bioreason, Inc.
Sante FE, NM

Pearson Education, Inc.
Upper Saddle River, NJ 07458

Library of Congress Cataloging-in-Publication Data on file

Vice President and Editorial Director, ECS: *Marcia J.Horton*
Executive Editor: *Eric Svendsen*
Associate Editor: *Dee Bernhard*
Vice President and Director of Production and Manufacturing, ESM: *David W. Riccardi*
Executive Managing Editor: *Vince O'Brien*
Managing Editor: *David A. George*
Production Editor: *Tamar Savir*
Director of Creative Services: *Paul Belfanti*
Creative Director: *Carole Anson*
Art Director: *Jayne Conte*
Art Editor: *Greg Dulles*
Manufacturing Manager: *Trudy Pisciotti*
Manufacturing Buyer: *Lisa McDowell*
Marketing Manager: *Holly Stark*

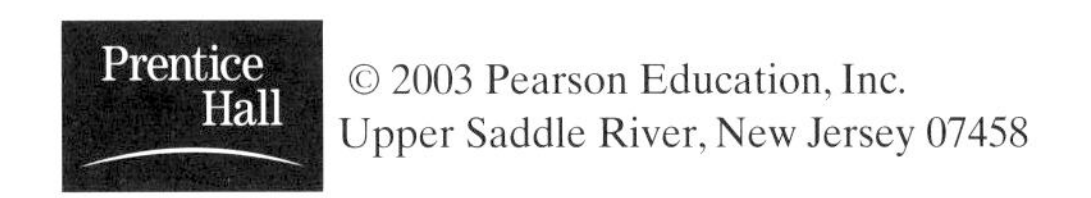

Upper Saddle River, New Jersey 07458

Printed in the United States of America.

10 9 8 7 6 5 4 3 2 1

ISBN 0-13-008175-2

Pearson Education Ltd., *London*
Pearson Education Australia Pty. Ltd., *Sydney*
Pearson Education Singapore, Pte. Ltd.
Pearson Education North Asia Ltd., *Hong Kong*
Pearson Education Canada, Inc., *Toronto*
Pearson Educación de Mexico, S.A. de C.V.
Pearson Education—Japan, *Tokyo*
Pearson Education Malaysia, Pte. Ltd.
Pearson Education, *Upper Saddle River, New Jersey*

About ESource

ESource—The Prentice Hall Engineering Source—www.prenhall.com/esource

ESource—The Prentice Hall Engineering Source gives professors the power to harness the full potential of their text and their first-year engineering course. More than just a collection of books, ESource is a unique publishing system revolving around the ESource website—www.prenhall.com/esource. ESource enables you to put your stamp on your book just as you do your course. It lets you:

Control You choose exactly what chapter or sections are in your book and in what order they appear. Of course, you can choose the entire book if you'd like and stay with the authors' original order.

Optimize Get the most from your book and your course. ESource lets you produce the optimal text for your students needs.

Customize You can add your own material anywhere in your text's presentation, and your final product will arrive at your bookstore as a professionally formatted text. Of course, all titles in this series are available as stand-alone texts, or as bundles of two or more books sold at a discount. Contact your PH sales rep for discount information.

ESource ACCESS

Professors who choose to bundle two or more texts from the ESource series for their class, or use an ESource custom book will be providing their students with complete access to the library of ESource content. All bundles and custom books will come with a student password that gives web ESource ACCESS to all information on the site. This passcode is free and is valid for one year after initial log-on. We've designed ESource ACCESS to provide students a flexible, searchable, on-line resource. Professors may also choose to deliver custom ESource content via the web only using ESource ACCESS passcodes. Contact your PH sales rep for more information.

ESource Content

All the content in ESource was written by educators specifically for freshman/first-year students. Authors tried to strike a balanced level of presentation, an approach that was neither formulaic nor trivial, and one that did not focus too heavily on advanced topics that most introductory students do not encounter until later classes. Because many professors do not have extensive time to cover these topics in the classroom, authors prepared each text with the idea that many students would use it for self-instruction and independent study. Students should be able to use this content to learn the software tool or subject on their own.

While authors had the freedom to write texts in a style appropriate to their particular subject, all followed certain guidelines created to promote a consistency that makes students comfortable. Namely, every chapter opens with a clear set of **Objectives**, includes **Practice Boxes** throughout the chapter, and ends with a number of **Problems**, and a list of **Key Terms**. **Applications Boxes** are spread throughout the book with the intent of giving students a real-world perspective of engineering. **Success Boxes** provide the student with advice about college study skills, and help students avoid the common pitfalls of first-year students. In addition, this series contains an entire book titled ***Engineering Success*** by Peter Schiavone of the University of Alberta intended to expose students quickly to what it takes to be an engineering student.

Creating Your Book

Using ESource is simple. You preview the content either on-line or through examination copies of the books you can request on-line, from your PH sales rep, or by calling 1-800-526-0485. Create an on-line outline of the content you want, in the order you want, using ESource's simple interface. Either type or cut and paste your own material and insert it into the text flow. You can preview the overall organization of the text you've created at anytime (please note, since this preview is immediate, it comes unformatted.), then press another button and receive an order number for your own custom book. If you are not ready to order, do nothing—ESource will save your work. You can come back at any time and change, re-arrange, or add more material to your creation. Once you're finished and you have an ISBN, give it to your bookstore and your book will arrive on their shelves four to six weeks after they order. Your custom desk copies with their instructor supplements will arrive at your address at the same time.

To learn more about this new system for creating the perfect textbook, go to www.prenhall.com/esource. You can either go through the on-line walkthrough of how to create a book, or experiment yourself.

Supplements

Adopters of ESource receive an instructor's CD that contains professor and student code from the books in the series, as well as other instruction aides provided by authors. The website also holds approximately **350 PowerPoint transparencies** created by Jack Leifer of University of Kentucky–Paducah available to download. Professors can either follow these transparencies as pre-prepared lectures or use them as the basis for their own custom presentations.

Titles in the ESource Series

Design Concepts for Engineers, 2/e
0-13-093430-5
Mark Horenstein

Engineering Success, 2/e
0-13-041827-7
Peter Schiavone

Engineering Design and Problem Solving, 2E
ISBN 0-13-093399-6
Steven K. Howell

Exploring Engineering
ISBN 0-13-093442-9
Joe King

Engineering Ethics
0-13-784224-4
Charles B. Fleddermann

Engineering Design—A Day in the Life of Four Engineers
0-13-085089-6
Mark N. Horenstein

Introduction to Engineering Analysis
0-13-016733-9
Kirk D. Hagen

Introduction to Engineering Experimentation
0-13-032835-9
Ronald W. Larsen, John T. Sears, and Royce Wilkinson

Introduction to Mechanical Engineering
0-13-019640-1
Robert Rizza

Introduction to Electrical and Computer Engineering
0-13-033363-8
Charles B. Fleddermann and Martin Bradshaw

Introduction to MATLAB 6
0-13-032845-6
Delores Etter and David C. Kuncicky, with Douglas W. Hull

Introduction to MATLAB
0-13-013149-0
Delores Etter with David C. Kuncicky

Introduction to Mathcad 2000
0-13-020007-7
Ronald W. Larsen

Introduction to Mathcad
0-13-937493-0
Ronald W. Larsen

Introduction to Maple
0-13-095133-1
David I. Schwartz

Mathematics Review
0-13-011501-0
Peter Schiavone

Power Programming with VBA/Excel
0-13-047377-4
Steven C. Chapra

Introduction to Excel 2002
0-13-008175-2
David C. Kuncicky

Introduction to Excel, 2/e
0-13-016881-5
David C. Kuncicky

Engineering with Excel
ISBN 0-13-017696-6
Ronald W. Larsen

Introduction to Word 2002
0-13-008170-1
David C. Kuncicky

Introduction to Word
0-13-254764-3
David C. Kuncicky

Introduction to PowerPoint 2002
0-13-008179-5
Jack Leifer

Introduction to PowerPoint
0-13-040214-1
Jack Leifer

Graphics Concepts
0-13-030687-8
Richard M. Lueptow

Graphics Concepts with SolidWorks
0-13-014155-0
Richard M. Lueptow and Michael Minbiole

Graphics Concepts with Pro/ENGINEER
0-13-014154-2
Richard M. Lueptow, Jim Steger, and Michael T. Snyder

Introduction to AutoCAD 2000
0-13-016732-0
Mark Dix and Paul Riley

Introduction to AutoCAD, R. 14
0-13-011001-9
Mark Dix and Paul Riley

Introduction to UNIX
0-13-095135-8
David I. Schwartz

Introduction to the Internet, 3/e
0-13-031355-6
Scott D. James

Introduction to Visual Basic 6.0
0-13-026813-5
David I. Schneider

Introduction to C
0-13-011854-0
Delores Etter

Introduction to C++
0-13-011855-9
Delores Etter

Introduction to FORTRAN 90
0-13-013146-6
Larry Nyhoff and Sanford Leestma

Introduction to Java
0-13-919416-9
Stephen J. Chapman

http://emissary.prenhall.com/esource/

About the Authors

No project could ever come to pass without a group of authors who have the vision and the courage to turn a stack of blank paper into a book. The authors in this series, who worked diligently to produce their books, provide the building blocks of the series.

Martin D. Bradshaw was born in Pittsburg, KS in 1936, grew up in Kansas and the surrounding states of Arkansas and Missouri, graduating from Newton High School, Newton, KS in 1954. He received the B.S.E.E. and M.S.E.E. degrees from the University of Wichita in 1958 and 1961, respectively. A Ford Foundation fellowship at Carnegie Institute of Technology followed from 1961 to 1963 and he received the Ph.D. degree in electrical engineering in 1964. He spent his entire academic career with the Department of Electrical and Computer Engineering at the University of New Mexico (1961-1963 and 1991-1996). He served as the Assistant Dean for Special Programs with the UNM College of Engineering from 1974 to 1976 and as the Associate Chairman for the EECE Department from 1993 to 1996. During the period 1987-1991 he was a consultant with his own company, EE Problem Solvers. During 1978 he spent a sabbatical year with the State Electricity Commission of Victoria, Melbourne, Australia. From 1979 to 1981 he served an IPA assignment as a Project Officer at the U.S. Air Force Weapons Laboratory, Kirkland AFB, Albuquerque, NM. He has won numerous local, regional, and national teaching awards, including the George Westinghouse Award from the ASEE in 1973. He was awarded the IEEE Centennial Medal in 2000.

Acknowledgments: Dr. Bradshaw would like to acknowledge his late mother, who gave him a great love of reading and learning, and his father, who taught him to persist until the job is finished. The encouragement of his wife, Jo, and his six children is a never-ending inspiration.

Stephen J. Chapman received a B.S. degree in Electrical Engineering from Louisiana State University (1975), the M.S.E. degree in Electrical Engineering from the University of Central Florida (1979), and pursued further graduate studies at Rice University. Mr. Chapman is currently Manager of Technical Systems for British Aerospace Australia, in Melbourne, Australia. In this position, he provides technical direction and design authority for the work of younger engineers within the company. He also continues to teach at local universities on a part-time basis.

Mr. Chapman is a Senior Member of the Institute of Electrical and Electronics Engineers (and several of its component societies). He is also a member of the Association for Computing Machinery and the Institution of Engineers (Australia).

Steven C. Chapra presently holds the Louis Berger Chair for Computing and Engineering in the Civil and Environmental Engineering Department at Tufts University. Dr. Chapra received engineering degrees from Manhattan College and the University of Michigan. Before joining the faculty at Tufts, he taught at Texas A&M University, the University of Colorado, and Imperial College, London. His research interests focus on surface water-quality modeling and advanced computer applications in environmental engineering. He has published over 50 refereed journal articles, 20 software packages and 6 books. He has received a number of awards including the 1987 ASEE Merriam/Wiley Distinguished Author Award, the 1993 Rudolph Hering Medal, and teaching awards from Texas A&M, the University of Colorado, and the Association of Environmental Engineering and Science Professors.

Acknowledgments: To the Berger Family for their many contributions to engineering education. I would also like to thank David Clough for his friendship and insights, John Walkenbach for his wonderful books, and my colleague Lee Minardi and my students Kenny William, Robert Viesca and Jennifer Edelmann for their suggestions.

Mark Dix began working with AutoCAD in 1985 as a programmer for CAD Support Associates, Inc. He helped design a system for creating estimates and bills of material directly from AutoCAD drawing databases for use in the automated conveyor industry. This system became the basis for systems still widely in use today. In 1986 he began collaborating with Paul Riley to create AutoCAD training materials, combining Riley's background in industrial design and training with Dix's background in writing, curriculum development, and programming. Mr. Dix received the M.S. degree in education from the University of Massachusetts. He is currently the Director of Dearborn Academy High School in Arlington, Massachusetts.

Delores M. Etter is a Professor of Electrical and Computer Engineering at the University of Colorado. Dr. Etter was a faculty member at the University of New Mexico and also a Visiting Professor at Stanford University. Dr. Etter was responsible for the Freshman Engineering Program at the University of New Mexico and is active in the Integrated Teaching Laboratory at the University of Colorado. She was elected a Fellow of the Institute of Electrical and Electronics Engineers for her contributions to education and for her technical leadership in digital signal processing.

Charles B. Fleddermann is a professor in the Department of Electrical and Computer Engineering at the University of New Mexico in Albuquerque, New Mexico. All of his degrees are in electrical engineering: his Bachelor's degree from the University of Notre Dame, and the Master's and Ph.D. from the University of Illinois at Urbana-Champaign. Prof. Fleddermann developed an engineering ethics course for his department in response to the ABET requirement to incorporate ethics topics into the undergraduate engineering curriculum. *Engineering Ethics* was written as a vehicle for presenting ethical theory, analysis, and problem solving to engineering undergraduates in a concise and readily accessible way.

Acknowledgments: I would like to thank Profs. Charles Harris and Michael Rabins of Texas A & M University whose NSF sponsored workshops on engineering ethics got me started thinking in this field. Special thanks to my wife Liz, who proofread the manuscript for this book, provided many useful suggestions, and who helped me learn how to teach "soft" topics to engineers.

Kirk D. Hagen is a professor at Weber State University in Ogden, Utah. He has taught introductory-level engineering courses and upper-division thermal science courses at WSU since 1993. He received his B.S. degree in physics from Weber State College and his M.S. degree in mechanical engineering from Utah State University, after which he worked as a thermal designer/analyst in the aerospace and electronics industries. After several years of engineering practice, he resumed his formal education, earning his Ph.D. in mechanical engineering at the University of Utah. Hagen is the author of an undergraduate heat transfer text.

Mark N. Horenstein is a Professor in the Department of Electrical and Computer Engineering at Boston University. He has degrees in Electrical Engineering from M.I.T. and U.C. Berkeley and has been involved in teaching engineering design for the greater part of his academic career. He devised and developed the senior design project class taken by all electrical and computer engineering students at Boston University. In this class, the students work for a virtual engineering company developing products and systems for real-world engineering and social-service clients.

Acknowledgments: I would like to thank Prof. James Bethune, the architect of the Peak Performance event at Boston University, for his permission to highlight the competition in my text. Several of the ideas relating to brainstorming and teamwork were derived from a

workshop on engineering design offered by Prof. Charles Lovas of Southern Methodist University. The principles of estimation were derived in part from a freshman engineering problem posed by Prof. Thomas Kincaid of Boston University.

Steven Howell is the Chairman and a Professor of Mechanical Engineering at Lawrence Technological University. Prior to joining LTU in 2001, Dr. Howell led a knowledge-based engineering project for Visteon Automotive Systems and taught computer-aided design classes for Ford Motor Company engineers. Dr. Howell also has a total of 15 years experience as an engineering faculty member at Northern Arizona University, the University of the Pacific, and the University of Zimbabwe. While at Northern Arizona University, he helped develop and implement an award-winning interdisciplinary series of design courses simulating a corporate engineering-design environment.

Douglas W. Hull is a graduate student in the Department of Mechanical Engineering at Carnegie Mellon University in Pittsburgh, Pennsylvania. He is the author of *Mastering Mechanics I Using Matlab 5*, and contributed to *Mechanics of Materials* by Bedford and Liechti. His research in the Sensor Based Planning lab involves motion planning for hyper-redundant manipulators, also known as serpentine robots.

Scott D. James is a staff lecturer at Kettering University (formerly GMI Engineering & Management Institute) in Flint, Michigan. He is currently pursuing a Ph.D. in Systems Engineering with an emphasis on software engineering and computer-integrated manufacturing. He chose teaching as a profession after several years in the computer industry. "I thought that it was really important to know what it was like outside of academia. I wanted to provide students with classes that were up to date and provide the information that is really used and needed."

Acknowledgments: Scott would like to acknowledge his family for the time to work on the text and his students and peers at Kettering who offered helpful critiques of the materials that eventually became the book.

Joe King received the B.S. and M.S. degrees from the University of California at Davis. He is a Professor of Computer Engineering at the University of the Pacific, Stockton, CA, where he teaches courses in digital design, computer design, artificial intelligence, and computer networking. Since joining the UOP faculty, Professor King has spent yearlong sabbaticals teaching in Zimbabwe, Singapore, and Finland. A licensed engineer in the state of California, King's industrial experience includes major design projects with Lawrence Livermore National Laboratory, as well as independent consulting projects. Prof. King has had a number of books published with titles including M*atlab*, MathCAD, Exploring Engineering, and Engineering and Society.

David C. Kuncicky is a native Floridian. He earned his Baccalaureate in psychology, Master's in computer science, and Ph.D. in computer science from Florida State University. He has served as a faculty member in the Department of Electrical Engineering at the FAMU–FSU College of Engineering and the Department of Computer Science at Florida State University. He has taught computer science and computer engineering courses for over 15 years. He has published research in the areas of intelligent hybrid systems and neural networks. He is currently the Director of Engineering at Bioreason, Inc. in Sante Fe, New Mexico.

Acknowledgments: Thanks to Steffie and Helen for putting up with my late nights and long weekends at the computer. Finally, thanks to Susan Bassett for having faith in my abilities, and for providing continued tutelage and support.

Ron Larsen is a Professor of Chemical Engineering at Montana State University, and received his Ph.D. from the Pennsylvania State University. He was initially attracted to engineering by the challenges the profession offers, but also appreciates that engineering is a serving profession. Some of the greatest challenges he has faced while teaching have involved non-traditional teaching methods, including evening courses for practicing engineers and teaching through an interpreter at the Mongolian National University. These experiences have provided tremendous opportunities to learn new ways to communicate technical material. Dr. Larsen views modern software as one of the new tools that will radically alter the way engineers work, and his book *Introduction to MathCAD* was written to help young engineers prepare to meet the challenges of an ever-changing workplace.

Acknowledgments: To my students at Montana State University who have endured the rough drafts and typos, and who still allow me to experiment with their classes—my sincere thanks.

Sanford Leestma is a Professor of Mathematics and Computer Science at Calvin College, and received his Ph.D. from New Mexico State University. He has been the long-time co-author of successful textbooks on Fortran, Pascal, and data structures in Pascal. His current research interest are in the areas of algorithms and numerical computation.

Jack Leifer is an Assistant Professor in the Department of Mechanical Engineering at the University of Kentucky Extended Campus Program in Paducah, and was previously with the Department of Mathematical Sciences and Engineering at the University of South Carolina–Aiken. He received his Ph.D. in Mechanical Engineering from the University of Texas at Austin in December 1995. His current research interests include the modeling of sensors for manufacturing, and the use of Artificial Neural Networks to predict corrosion.

Acknowledgments: I'd like to thank my colleagues at USC–Aiken, especially Professors Mike May and Laurene Fausett, for their encouragement and feedback; and my parents, Felice and Morton Leifer, for being there and providing support (as always) as I completed this book.

Richard M. Lueptow is the Charles Deering McCormick Professor of Teaching Excellence and Associate Professor of Mechanical Engineering at Northwestern University. He is a native of Wisconsin and received his doctorate from the Massachusetts Institute of Technology in 1986. He teaches design, fluid mechanics, an spectral analysis techniques. Rich has an active research program on rotating filtration, Taylor Couette flow, granular flow, fire suppression, and acoustics. He has five patents and over 40 refereed journal and proceedings papers along with many other articles, abstracts, and presentations.

Acknowledgments: Thanks to my talented and hard-working co-authors as well as the many colleagues and students who took the tutorial for a "test drive." Special thanks to Mike Minbiole for his major contributions to Graphics Concepts with SolidWorks. Thanks also to Northwestern University for the time to work on a book. Most of all, thanks to my loving wife, Maiya, and my children, Hannah and Kyle, for supporting me in this endeavor. (Photo courtesy of Evanston Photographic Studios, Inc.)

Larry Nyhoff is a Professor of Mathematics and Computer Science at Calvin College. After doing bachelor's work at Calvin, and Master's work at Michigan, he received a Ph.D. from Michigan State and also did graduate work in computer science at Western Michigan. Dr. Nyhoff has taught at Calvin for the past 34 years—mathematics at first and computer science for the past several years.

Acknowledgments: We thank our families—Shar, Jeff, Dawn, Rebecca, Megan, Sara, Greg, Julie, Joshua, Derek, Tom, Joan; Marge, Michelle, Sandy, Lory, Michael—for being patient and understanding. We thank God for allowing us to write this text.

Paul Riley is an author, instructor, and designer specializing in graphics and design for multimedia. He is a founding partner of CAD Support Associates, a contract service and professional training organization for computer-aided design. His 15 years of business experience and 20 years of teaching experience are supported by degrees in education and computer science. Paul has taught AutoCAD at the University of Massachusetts at Lowell and is presently teaching AutoCAD at Mt. Ida College in Newton, Massachusetts. He has developed a program,

Computer-aided Design for Professionals that is highly regarded by corporate clients and has been an ongoing success since 1982.

Robert Rizza is an Assistant Professor of Mechanical Engineering at North Dakota State University, where he teaches courses in mechanics and computer-aided design. A native of Chicago, he received the Ph.D. degree from the Illinois Institute of Technology. He is also the author of *Getting Started with Pro/ENGINEER*. Dr. Rizza has worked on a diverse range of engineering projects including projects from the railroad, bioengineering, and aerospace industries. His current research interests include the fracture of composite materials, repair of cracked aircraft components, and loosening of prostheses.

Peter Schiavone is a professor and student advisor in the Department of Mechanical Engineering at the University of Alberta, Canada. He received his Ph.D. from the University of Strathclyde, U.K. in 1988. He has authored several books in the area of student academic success as well as numerous papers in international scientific research journals. Dr. Schiavone has worked in private industry in several different areas of engineering including aerospace and systems engineering. He founded the first Mathematics Resource Center at the University of Alberta, a unit designed specifically to teach new students the necessary *survival skills* in mathematics and the physical sciences required for success in first-year engineering. This led to the Students' Union Gold Key Award for outstanding contributions to the university. Dr. Schiavone lectures regularly to freshman engineering students and to new engineering professors on engineering success, in particular about maximizing students' academic performance.

Acknowledgements: Thanks to Richard Felder for being such an inspiration; to my wife Linda for sharing my dreams and believing in me; and to Francesca and Antonio for putting up with Dad when working on the text.

David I. Schneider holds an A.B. degree from Oberlin College and a Ph.D. degree in Mathematics from MIT. He has taught for 34 years, primarily at the University of Maryland. Dr. Schneider has authored 28 books, with one-half of them computer programming books. He has developed three customized software packages that are supplied as supplements to over 55 mathematics textbooks. His involvement with computers dates back to 1962, when he programmed a special purpose computer at MIT's Lincoln Laboratory to correct errors in a communications system.

David I. Schwartz is an Assistant Professor in the Computer Science Department at Cornell University and earned his B.S., M.S., and Ph.D. degrees in Civil Engineering from State University of New York at Buffalo. Throughout his graduate studies, Schwartz combined principles of computer science to applications of civil engineering. He became interested in helping students learn how to apply software tools for solving a variety of engineering problems. He teaches his students to learn incrementally and practice frequently to gain the maturity to tackle other subjects. In his spare time, Schwartz plays drums in a variety of bands.

Acknowledgments: I dedicate my books to my family, friends, and students who all helped in so many ways.

Many thanks go to the schools of Civil Engineering and Engineering & Applied Science at State University of New York at Buffalo where I originally developed and tested my UNIX and Maple books. I greatly appreciate the opportunity to explore my goals and all the help from everyone at the Computer Science Department at Cornell.

John T. Sears received the Ph.D. degree from Princeton University. Currently, he is a Professor and the head of the Department of Chemical Engineering at Montana State University. After leaving Princeton he worked in research at Brookhaven National Laboratory and Esso Research and Engineering, until he took a position at West Virginia University. He came to MSU in 1982, where he has served as the Director of the College of Engineering Minority Program and Interim Director for BioFilm Engineering. Prof. Sears has written a book on air pollution and economic development, and over 45 articles in engineering and engineering education.

Michael T. Snyder is President of Internet startup Appointments123.com. He is a native of Chicago, and he received his Bachelor of Science degree in Mechanical Engineering from the University of Notre Dame. Mike also graduated with honors from Northwestern University's Kellogg Graduate School of Management in 1999 with his Masters of Management degree. Before Appointments123.com, Mike was a mechanical engineer in new product development for Motorola Cellular and Acco Office Products. He has received four patents for his mechanical design work. "Pro/ENGINEER was an invaluable design tool for me, and I am glad to help students learn the basics of Pro/ENGINEER."

Acknowledgments: Thanks to Rich Lueptow and Jim Steger for inviting me to be a part of this great project. Of course, thanks to my wife Gretchen for her support in my various projects.

Jim Steger is currently Chief Technical Officer and cofounder of an Internet applications company. He graduated with a Bachelor of Science degree in Mechanical Engineering from Northwestern University. His prior work included mechanical engineering assignments at Motorola and Acco Brands. At Motorola, Jim worked on part design for two-way radios and was one of the lead mechanical engineers on a cellular phone product line. At Acco Brands, Jim was the sole engineer on numerous office product designs. His Worx stapler has won design awards in the United States and in Europe. Jim has been a Pro/ENGINEER user for over six years.

Acknowledgments: Many thanks to my co-authors, especially Rich Lueptow for his leadership on this project. I would also like to thank my family for their continuous support.

Royce Wilkinson received his undergraduate degree in chemistry from Rose-Hulman Institute of Technology in 1991 and the Ph.D. degree in chemistry from Montana State University in 1998 with research in natural product isolation from fungi. He currently resides in Bozeman, MT and is involved in HIV drug research. His research interests center on biological molecules and their interactions in the search for pharmaceutical advances.

Reviewers

ESource benefited from a wealth of reviewers who on the series from its initial idea stage to its completion. Reviewers read manuscripts and contributed insightful comments that helped the authors write great books. We would like to thank everyone who helped us with this project.

Concept Document

Naeem Abdurrahman *University of Texas, Austin*
Grant Baker *University of Alaska, Anchorage*
Betty Barr *University of Houston*
William Beckwith *Clemson University*
Ramzi Bualuan *University of Notre Dame*
Dale Calkins *University of Washington*
Arthur Clausing *University of Illinois at Urbana–Champaign*
John Glover *University of Houston*
A.S. Hodel *Auburn University*
Denise Jackson *University of Tennessee, Knoxville*
Kathleen Kitto *Western Washington University*
Terry Kohutek *Texas A&M University*
Larry Richards *University of Virginia*
Avi Singhal *Arizona State University*
Joseph Wujek *University of California, Berkeley*
Mandochehr Zoghi *University of Dayton*

Books

Naeem Abdurrahman *University of Texas, Austin*
Stephen Allan *Utah State University*
Anil Bajaj *Purdue University*
Grant Baker *University of Alaska–Anchorage*
William Beckwith *Clemson University*
Haym Benaroya *Rutgers University*
John Biddle *California State Polytechnic University*
Tom Bledsaw *ITT Technical Institute*
Fred Boadu *Duke University*
Tom Bryson *University of Missouri, Rolla*
Ramzi Bualuan *University of Notre Dame*
Dan Budny *Purdue University*
Betty Burr *University of Houston*
Dale Calkins *University of Washington*
Harish Cherukuri *University of North Carolina –Charlotte*
Arthur Clausing *University of Illinois*
Barry Crittendon *Virginia Polytechnic and State University*
James Devine *University of South Florida*
Ron Eaglin *University of Central Florida*
Dale Elifrits *University of Missouri, Rolla*
Patrick Fitzhorn *Colorado State University*
Susan Freeman *Northeastern University*
Frank Gerlitz *Washtenaw College*
Frank Gerlitz *Washtenaw Community College*
John Glover *University of Houston*
John Graham *University of North Carolina–Charlotte*
Ashish Gupta *SUNY at Buffalo*
Otto Gygax *Oregon State University*
Malcom Heimer *Florida International University*
Donald Herling *Oregon State University*
Thomas Hill *SUNY at Buffalo*
A.S. Hodel *Auburn University*
James N. Jensen *SUNY at Buffalo*
Vern Johnson *University of Arizona*
Autar Kaw *University of South Florida*
Kathleen Kitto *Western Washington University*
Kenneth Klika *University of Akron*
Terry L. Kohutek *Texas A&M University*
Melvin J. Maron *University of Louisville*
Robert Montgomery *Purdue University*
Mark Nagurka *Marquette University*
Romarathnam Narasimhan *University of Miami*
Soronadi Nnaji *Florida A&M University*
Sheila O'Connor *Wichita State University*
Michael Peshkin *Northwestern University*
Dr. John Ray *University of Memphis*
Larry Richards *University of Virginia*
Marc H. Richman *Brown University*
Randy Shih *Oregon Institute of Technology*
Avi Singhal *Arizona State University*
Tim Sykes *Houston Community College*
Neil R. Thompson *University of Waterloo*
Dr. Raman Menon Unnikrishnan *Rochester Institute of Technology*
Michael S. Wells *Tennessee Tech University*
Joseph Wujek *University of California, Berkeley*
Edward Young *University of South Carolina*
Garry Young *Oklahoma State University*
Mandochehr Zoghi *University of Dayton*

Contents

http://emissary.prenhall.com/esource/

1

Microsoft Excel Basics

1.1 INTRODUCTION TO WORKSHEETS

A *spreadsheet* is a rectangular grid composed of addressable units called *cells*. A cell is addressed by referencing its column number and row number. A cell may contain numerical data, textual data, macros, or formulas.

Spreadsheet application programs were originally intended to be used for financial calculations. The original electronic spreadsheets resembled the paper spreadsheets of an accountant. One characteristic of electronic spreadsheets that gives them an advantage over their paper counterparts is their ability to automatically recalculate all dependent values whenever a parameter is changed.

Over time, more and more functionality has been added to spreadsheet application programs. A variety of mathematical and engineering functions now exist within Excel. A number of analytical tools are also available, including scientific and engineering tools, statistical tools, data-mapping tools, and financial-analysis tools. Auxiliary functions includes graphing capability, database functions, and the ability to access the World Wide Web.

SECTION

OBJECTIVES

After reading this chapter, you should be able to:

- Describe how spreadsheets are used by engineers.
- Identify the main components on the Excel screen.
- Name at least four ways to get on-line help for Excel.
- Create and save a new worksheet.
- Open and edit an existing worksheet.
- Undo mistakes.
- Perform spelling and grammar checks on text items.
- Preview and print a worksheet.

As an engineering student, you may find that an advanced spreadsheet program such as Microsoft Excel will suffice for most of your computational and presentation needs. For example, Excel may be used to create technical reports. However, if you wish to create large formatted documents, you may want to use a document preparation tool such as Microsoft Word. Tables and charts may be easily exported from Excel into Word.

Excel also has some capability for database management. However, if you wish to manage large or sophisticated databases, a specialized database application such as Microsoft Access or Oracle is preferable.

In addition, Excel has fairly sophisticated mechanisms for performing mathematical and scientific analyses. For example, you can use the Analysis Toolpack in Excel to perform mathematical analysis. If the analysis is large or very sophisticated, however, you may want to use a specialized mathematical or matrix package such as MathCAD or MATLAB®.

The same principles hold for graphing (Harvard Graphics) or statistical analysis (SPSS). Excel is a general tool that performs many functions for small- to medium-sized problems. As the size or sophistication of the function increases, other tools may be more applicable.

Microsoft Excel uses the term *worksheet* to denote a spreadsheet. A worksheet can contain more types of items than a traditional paper spreadsheet. These include charts, links to Web pages, Visual Basic programs, and macros. We will treat the terms *worksheet* and *spreadsheet* synonymously in this text. Worksheets stored in a single file are called a *workbook*.

1.2 HOW TO USE THIS BOOK

This book is intended to get you, the engineering student, up and running with Excel 2002 as quickly as possible. Examples are geared towards engineering and mathematical problems. Try to read the book while sitting in front of a computer. Learn to use Excel by re-creating each example in the text. Perform the instructions in the boxes labeled **PRACTICE**.

The book is not intended to be a complete reference manual for Excel. It is much too short for that purpose. Many books on the market are more appropriate for use as complete reference manuals. However, if you are sitting at the computer, one of the best reference manuals is at your fingertips. The on-line Excel help tools provide an excellent resource if properly used. These help tools are described later in this chapter.

1.3 TYPOGRAPHIC CONVENTIONS USED IN THIS BOOK

Throughout the text, the following conventions will be used:

1.3.1 Selection with the Mouse

The book frequently asks you to move the mouse cursor over a particular item and then click and release the left mouse button. This action is repeated so many times in the text that it will be abbreviated as follows:

Choose **Item**.

If the mouse button is not to be released or if the right mouse button is to be used, then this will be stated explicitly.

A button, icon, or menu item, you are to select with the mouse will be printed in boldface font. A key you should press will also be printed in boldface font. For example, if you are asked to choose an item from the menu at the top of the screen labeled *File*, then it will be written as follows:

Choose **File** from the Menu bar.

1.3.2 Multiple Selections

The book frequently refers to selections that require more than one step. For example, to see a print preview, perform the following steps:

Step 1: Choose **File** from the Menu bar (at the top of the screen).

Step 2: Choose **Print Preview** from the drop-down menu that appears.

Multiple selections such as this will be abbreviated by separating choices with a right arrow. For example, the two steps listed will be abbreviated as follows:

Choose File → Print Preview from the Menu bar.

1.3.3 Multiple Keystrokes

If you are asked to simultaneously press multiple keys, the key names will be printed in bold font and will be separated with a plus sign. For example, to undo a typing change, you can simultaneously press the **Ctrl** key and the **Z** key. This will be abbreviated as follows:

To undo typing, press **Ctrl + Z.**

1.3.4 Literal Expressions

A word or phrase that is a literal transcription will be printed in italics. For example, the title bar at the top of the screen should contain the text *Microsoft Excel*. Another example is the literal name of a box or menu item, such as the following:

Check the box labeled *Equal To*.

1.3.5 Key Terms

The first time a key term is used, it will be italicized. Key terms are summarized at the end of each chapter.

1.4 UNDERSTANDING THE EXCEL 2002 SCREEN

This section introduces you to the Microsoft Excel screen. To start the Excel program, place the cursor over the Excel icon and double click the left mouse button. The icon resembles Figure 1.1. The icon may be on your desktop or may be accessed from the Start menu.

A screen that resembles Figure 1.2 will appear. If the Tip of the Day box appears, close it for now by choosing **Close**. We will return to the Tip of the Day later in this chapter.

We'll now discuss each of the components on the screen. The Excel screen consists of a number of components, including

Figure 1.1. The Excel 2002 shortcut icon.

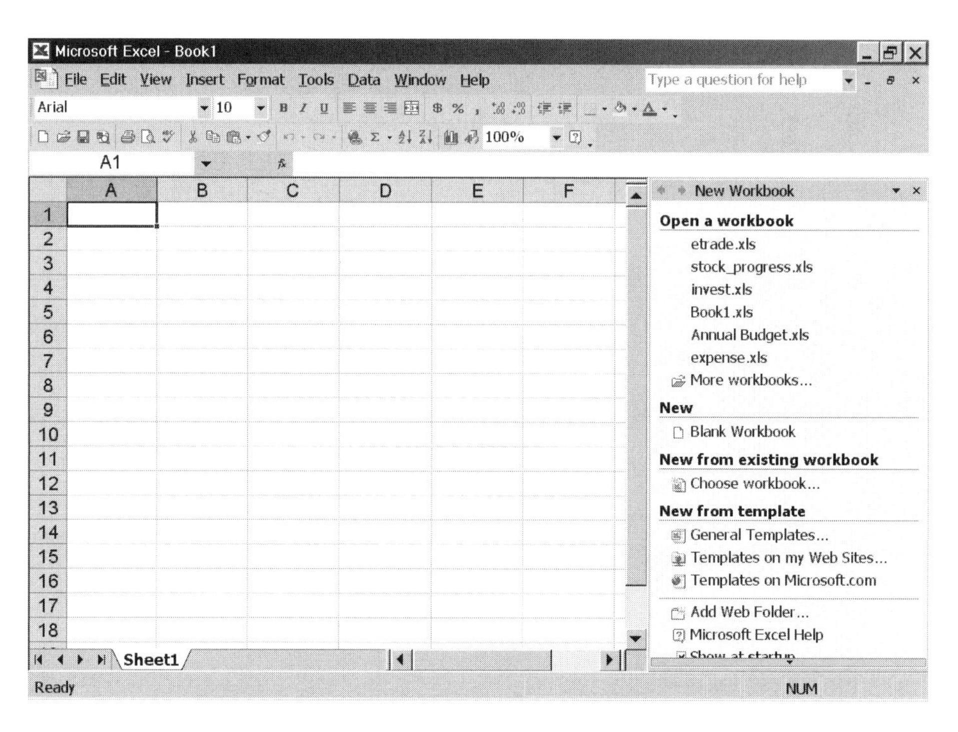

Figure 1.2. The Excel screen.

- the Title bar
- the Menu bar
- various toolbars
- the Formula bar
- the Workbook window
- the Task pane
- Sheet tabs
- scrollbars
- the Status bar

Try to become familiar with the names of these components as we proceed, as we will use these names throughout the book. Working generally from top to bottom, each of the components will be discussed in turn.

1.4.1 Title Bar

The bar at the top of the screen is called the *Title bar*. The Title bar contains the Excel icon and the name of the worksheet currently being edited. Since we did not specifically open a worksheet, Excel supplies a default name. In our example, the default name is *Book1*.

Figure 1.3 shows an example Title bar. You should see several buttons on the right-hand side of the Title bar. These are used to manipulate the window and will be discussed later in this chapter, in the section titled *Manipulating Windows*.

Figure 1.3. The Title bar.

Figure 1.4. The Menu bar.

1.4.2 Menu Bar

The next bar, which is just below the Title bar, is called the *Menu bar*. The Menu bar contains a list of menus. If you place the cursor over an item on the Menu bar and click the left mouse button, then a drop-down menu will appear. The drop-down menu for the *Window* menu item is shown in Figure 1.4. If the drop-down menu is not fully expanded, click on the small down arrow at the bottom of the drop-down menu.

When you place the cursor over an item, the item should change color. While holding the cursor over an item, click the left mouse button to execute the item. Try the following steps:

Step 1: Place the cursor over the **Window** menu item on the Menu bar.

Step 2: When the drop-down menu appears, drag the mouse and click on **Arrange**.

Recall that we will abbreviate these instructions as follows:

Choose **Window** → **Arrange** from the Menu bar.

1.4.3 Toolbars

The next several rows of icons on the Excel screen are called *toolbars*. A toolbar is a group of buttons related to a particular topic (e.g., drawing). The use of buttons on toolbars is an alternate method of executing commands. Most (but not all) of the commands that can be executed from a toolbar can also be executed from the Menu bar. There are more than a dozen toolbars for various functions.

We will discuss one toolbar now—the Standard toolbar. Some of the other toolbars will be discussed throughout the book.

Figure 1.5 shows the Standard toolbar. The toolbars on your screen may not match the example exactly. This is because the location and presence of toolbars may be customized. If all of the toolbars were displayed at once on the screen, there would not be much room left for anything else!

The Standard toolbar contains some of the most frequently used commands. If this toolbar does not appear on your screen, then choose **View** → **Toolbars** from the Menu bar. Make sure that the box labeled *Standard* is checked.

Move the mouse over the first icon on the Standard toolbar. Hold the mouse still for a few seconds, but don't click. A small window will appear that shows the name of the

Figure 1.5. The Standard toolbar.

Figure 1.6. Example of Tool Tip.

icon. This is called a *Tool Tip*. The Tool Tip for the first icon on the Standard toolbar is shown in Figure 1.6. The name of the first icon is *New*.

Clicking the **New** icon on the Standard toolbar has an effect similar to that of choosing **File** → **New** from the Menu bar. These are two methods for creating a new workbook.

PRACTICE!

Move the mouse slowly over the second icon on the Standard toolbar and view its Tool Tip. Try to locate the equivalently named command on the Menu bar. Do this for each of the icons on the Standard toolbar.

1.4.4 Customizing Toolbars

The choice, location, and composition of toolbars may be customized. To add or remove toolbars from the screen, choose **View** → **Toolbars** from the Menu bar. A list of toolbars will appear on your screen, as shown in Figure 1.7. From this menu, you can select or deselect any toolbars that you would like to have displayed on the screen.

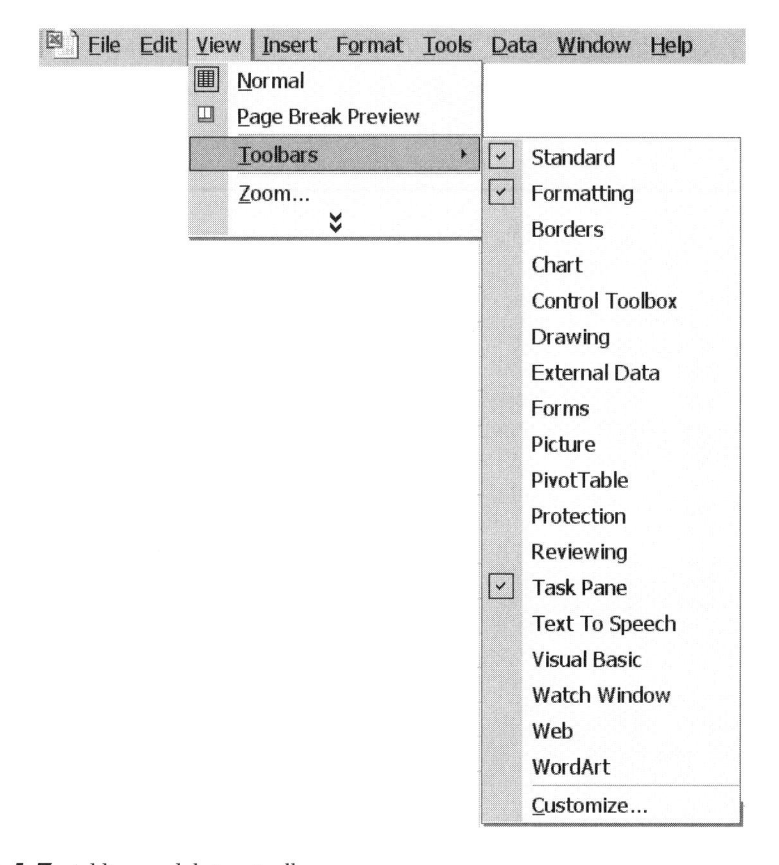

Figure 1.7. Adding or deleting toolbars.

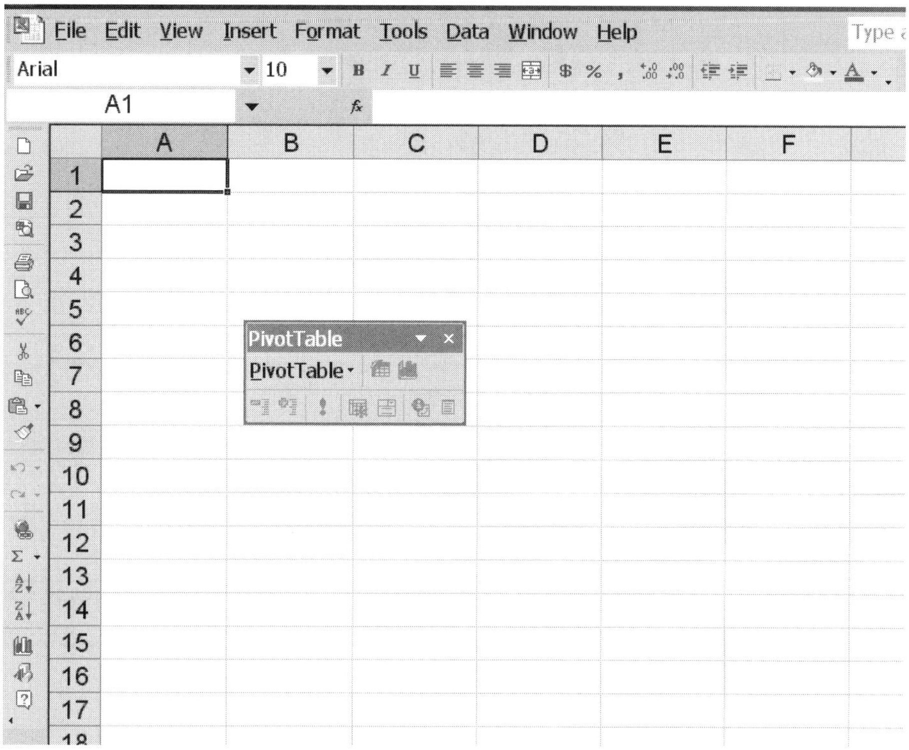

Figure 1.8. Example of customized toolbars.

Once a toolbar appears on the screen, you can move it by dragging the toolbar with the mouse. This is accomplished by moving the cursor to the very left side of the toolbar. There is a grey vertical line on the left side of each toolbar. Now, drag the toolbar to a new location. Toolbars may be placed along the left-hand side of the screen or in a separate box anywhere on the screen.

Figure 1.8 shows the Standard toolbar placed along the left side of the screen, the Formatting toolbar placed at the top of the screen, and the PivotTable toolbar placed near the center of the screen.

You can add or remove buttons from a toolbar by clicking on the small down arrow on the right end of the toolbar. A small drop-down menu will appear. Choose **Add or Remove Buttons** from the drop-down menu.

1.4.5 Formula Bar

The *Formula bar* displays the formula or constant value for a selected cell. Formulas may be typed into the formula window, or a built-in function may be chosen from a drop-down list.

Figure 1.9 shows the Formula bar. The left window of the Formula bar shows the target cell. This is where the results of the formula will be displayed. The formula itself is shown in the Formula window.

Choose the **Insert Function** icon, the Insert Function dialog box will appear, as shown in Figure 1.10. From the Insert Function dialog box, you can choose a function category and function name. In the example, we have chosen the category **Math & Trig** and the function **SIN**.

At the bottom of the Insert Function dialog box, a brief description of the function is displayed. The dialog box also has a search feature to help you locate a function. There are over 200 built-in functions available in Excel 2002.

Figure 1.9. The Formula bar.

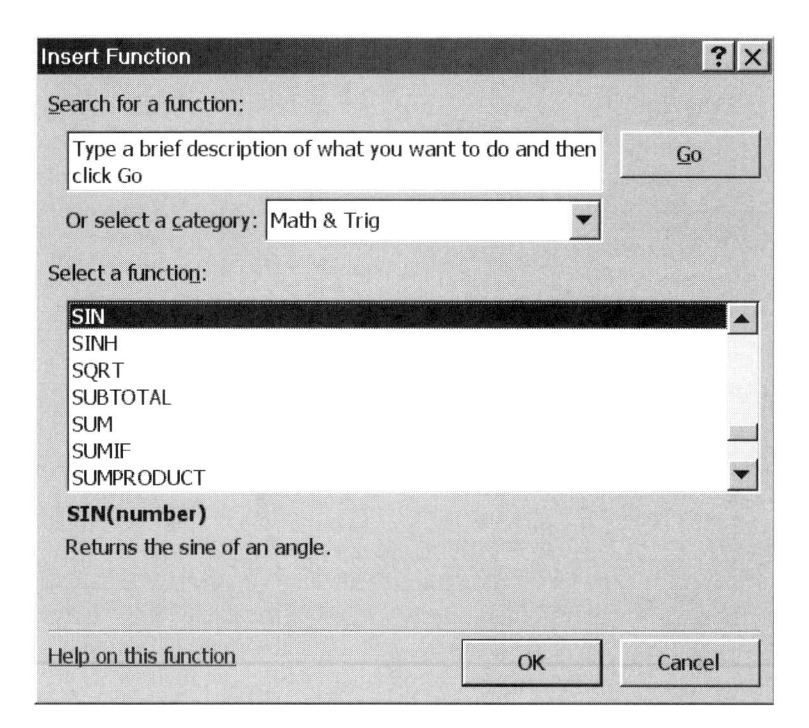

Figure 1.10. The Insert Function dialog box.

Choose the **SIN** function, then click **OK**. The Function Arguments dialog box will appear, as shown in Figure 1.11. The dialog prompts for the arguments to the named function.

A short explanation about the expected arguments appears in the bottom of the window. In this case, the **SIN** function takes its arguments in radians. The formula for converting radians to degrees is also displayed.

The arguments may be a range of cells, numbers, or other functions. We will discuss the formula bar in more detail in Chapter 3. For now, type

pi( ) / 2

as the argument. The effect of this is to call another built-in function named *pi* and divide the results by 2. Then click **OK**. The Function Arguments dialog box will disappear and the result will be displayed in the target cell.

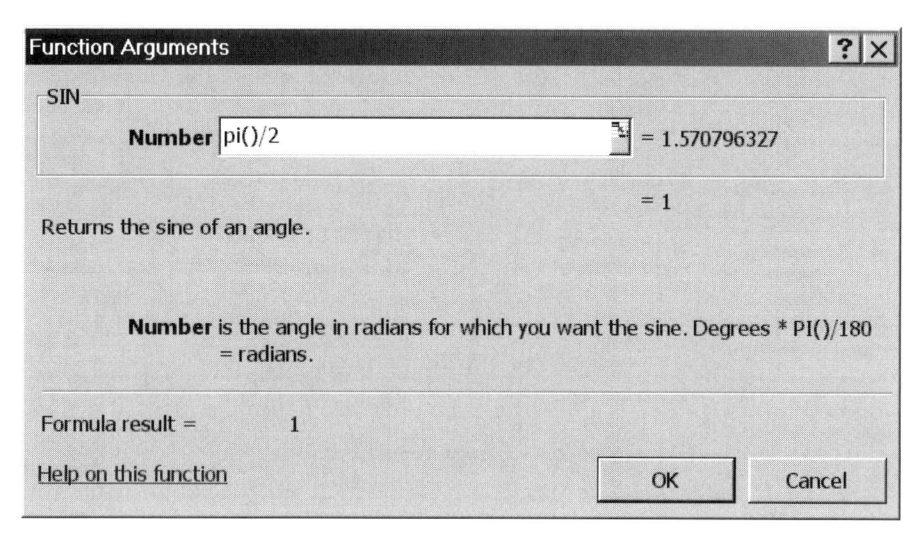

Figure 1.11. The Function Arguments dialog box.

1.4.6 Task Pane

The *Task pane* is a subwindow that holds commonly used commands. If the Task pane is not visible on your screen, choose **View** → **Task Pane** from the Menu bar. The Task pane may take several forms, but the most common is the New Document Task pane, which is shown in Figure 1.12. From the New Document Task pane you can open recently used workbooks or choose a template for a new workbook. We'll show you how to do this later in the chapter.

Figure 1.12. The New Document Task pane.

1.4.7 Workbook Window

The Workbook window is the area on the screen where data are entered and displayed. The Workbook window contains one or more worksheets.

The maximum size for a worksheet is 65,536 rows by 256 columns. The columns are labeled A, B, C, . . . AA, AB, . . . IV and the rows are labeled 1, 2, 3, . . . 65,536.

A single cell can be selected by placing the mouse over the cell and clicking the mouse. A range of cells can be selected by holding the left mouse button down and dragging it over the selected range. An entire column can be selected by clicking the left mouse button on the column label. An entire row can be selected by clicking on the row label. The entire worksheet can be selected by clicking on the blank box in the top left corner of the Workbook window. Figure 1.13 shows a Workbook window with column C selected.

1.4.8 Sheet Tabs

The Sheet Tab bar is positioned at the bottom of the screen. You can have more than one worksheet in a workbook. The Sheet Tab bar lists the worksheets in the workbook.

You can move quickly from sheet to sheet by selecting a Sheet tab. If there are more sheets that are visible on the Sheet Tab bar, then you can use the arrows to the left of the sheet tabs to move from sheet to sheet. By default, Excel creates a single worksheet when you create a new workbook. The maximum number of worksheets open at the same time in a workbook is limited only by the amount of memory on your computer. Figure 1.14 shows the Sheet Tab bar. We'll show you how to create new worksheets later in this chapter.

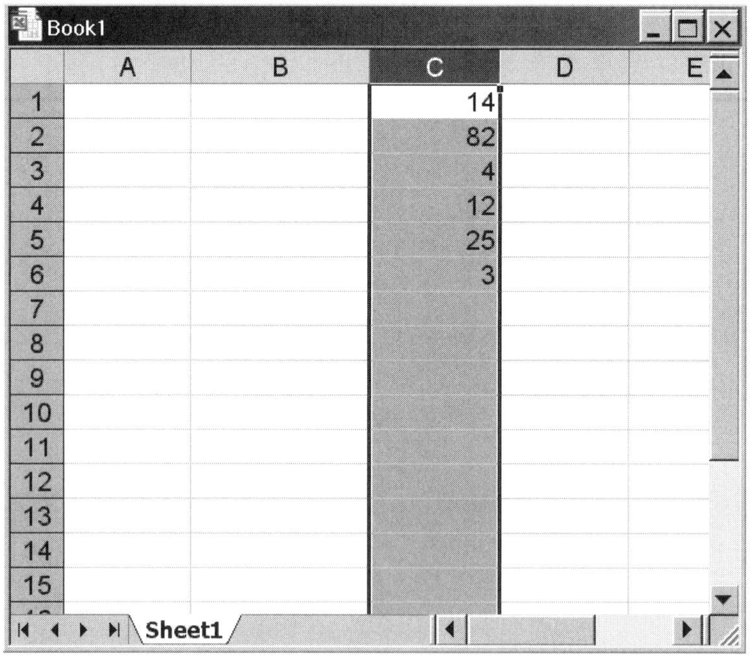

Figure 1.13. The Workbook window.

Figure 1.14. The Sheet Tab bar.

1.4.9 Scrollbars

The Vertical and Horizontal *scrollbars* are located along the right-hand side and bottom of the Workbook window. Many worksheets are much larger than the visible window. A scrollbar makes it possible to move quickly to any position on a worksheet.

You can move around the worksheet by dragging a scrollbar, or you can click on the arrows at either end of a scrollbar. Figure 1.15 shows the Horizontal and Vertical scrollbars.

1.4.10 Status Bar

The Status bar is normally positioned at the very bottom of the Excel screen. The *Status bar* displays information about a command in progress and displays the status of certain keys such as Num Lock, Caps Lock, and Scroll Lock. The Status bar depicted in Figure 1.16 shows that the Num Lock key is turned on and that Excel is ready for you to type something. If the status bar is not visible on your screen and you want it to be, choose **View** → **Status Bar** from the Menu bar.

Figure 1.15. The Horizontal and Vertical scrollbars.

Figure 1.16. The Status bar.

1.5 GETTING HELP

Excel contains a large on-line help system. To access the help menu, choose **Help** from the Menu bar. There are at least eight ways to obtain help. These include the following:

- choosing a section from the table of contents
- using the Answer Wizard
- using the Help Index
- using the Question panel on the Standard toolbar
- using the Office Assistant
- using the What's This feature
- learning from the Tip of the Day
- accessing help from the World Wide Web

Each of these methods will be discussed in the next sections.

1.5.1 Choosing from the Table of Contents

This method is useful if you have time to read about a general topic. Reading through the topic could serve as a tutorial, which is not the method to use if you have a specific question and you want an immediate answer. To access the Table of Contents,

Step 1: Make sure that the Office Assistant is turned off by choosing **Help → Hide the Office Assistant** from the Menu bar.

Step 2: Choose **Help → Microsoft Excel Help** from the Menu bar. Alternatively, you can press the **F1** key.

After you complete the preceding two steps, the Help dialog box will appear, as shown in Figure 1.17.

Step 1: Choose the Contents tab. The contents chapters can be expanded or compressed by clicking on the plus sign in front of each topic.

Step 2: A topic will have a question mark in front of it. When you click on a topic, the results will be displayed in the pane on the right. Figure 1.17 shows the results of clicking on a topic labeled *About collecting and pasting multiple items*.

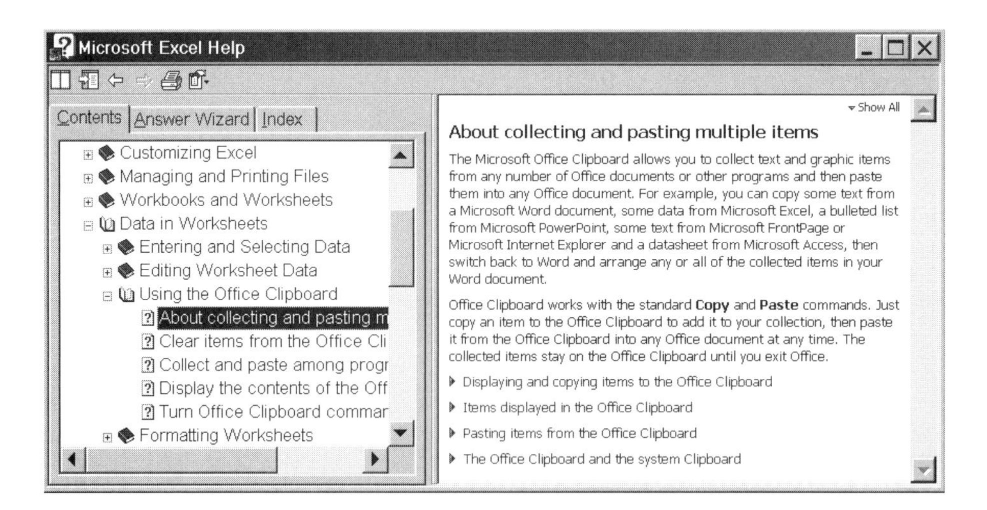

Figure 1.17. Using the Contents tab of the Help dialog box.

1.5.2 Using the Answer Wizard

The Answer Wizard allows you to ask questions. You can access the Answer Wizard by using the following steps:

Step 1: Make sure that the Office Assistant is turned off by choosing **Help → Hide the Office Assistant** from the Menu bar.

Step 2: Choose **Help → Microsoft Excel Help** from the Menu bar. Alternatively, you can press the **F1** key.

After you complete the preceding two steps, the Help dialog box will appear. Choose the **Answer Wizard** tab, and procede by using the following steps:

Step 1: Type a question in the box labeled *What would you like to.*

Step 2: Select a topic from the box labeled *Select a topic to.*

The results will appear in the pane on the right side of the Help dialog box. Figure 1.18 shows an example asking the question *How can I get help?*

1.5.3 Using the Help Index

The Help index feature allows you to access topics by using keywords. This is a rapid way to see all topics that contain a keyword. To use the Help index do the following:

Step 1: Make sure that the Office Assistant is turned off by choosing **Help → Hide the Office Assistant** from the Menu bar.

Step 2: Choose **Help → Microsoft Excel Help** from the Menu bar. Alternatively, you can press the **F1** key.

After you completed the preceding two steps, the Help dialog box will appear. Choose the **Index** tab, and procede by using the following steps:

Step 1: Type a keyword in the box labeled *Type keywords*, and then choose **Search**.

Step 2: Or choose a keyword from the scroll list labeled *Or choose keywords.*

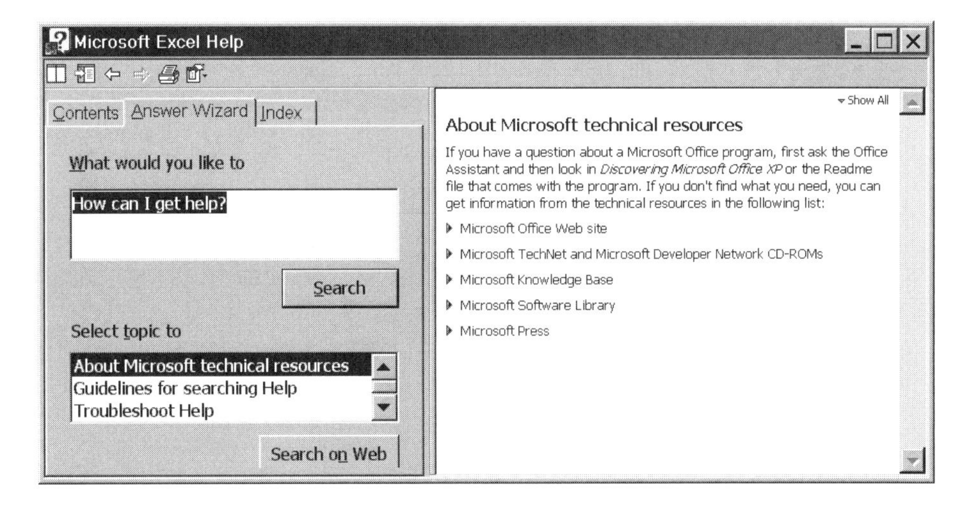

Figure 1.18. Using the Answer Wizard.

The results will appear in the pane on the right side of the Help dialog box. Figure 1.19 shows an example using the keywords *shortcut key.*

1.5.4 Using the Question Panel on the Standard Toolbar

A quick way to ask a question or type a keyword is to use the Question panel on the right side of the Standard toolbar. To use the Question panel,

Step 1: Type a keyword or question into the Question panel.

Step 2: Press **Enter**.

A drop-down menu of topics will appear, as shown in Figure 1.20. Click on a topic. The Help dialog box will appear and the chosen topic will be displayed.

1.5.5 Using the Office Assistant

The Office Assistant is intended to interactively guide you through many tasks. The Office Assistant has several personalities, including an animated paper clip and an animated Albert Einstein. To turn on the Office Assistant, choose **Help → Show the Office Assistant** from the Menu bar. The Office Assistant will then appear and will try to anticipate your help needs. To change Office Assistant options,

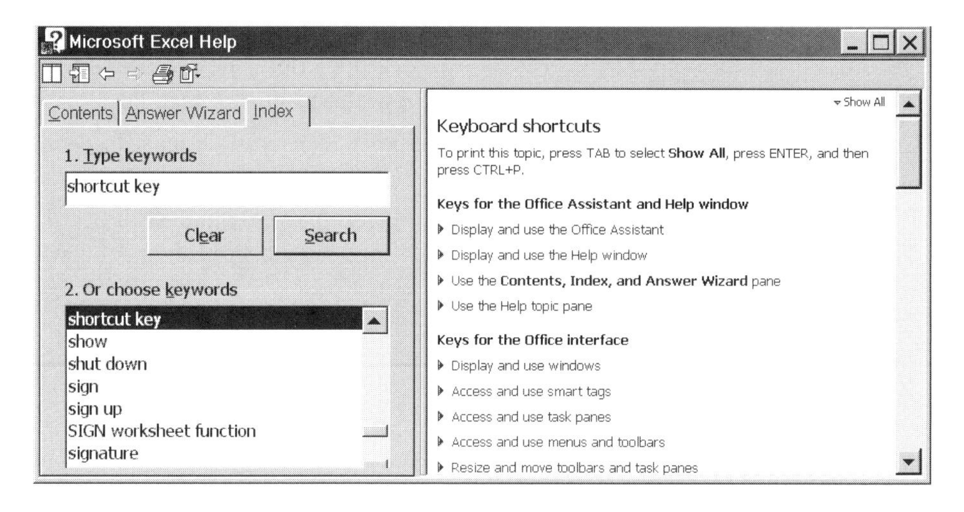

Figure 1.19. Using the Help Index feature.

Figure 1.20. Using the Question panel.

Step 1: Click on the Office Assistant.

Step 2: Choose **Options**.

The Office Assistant dialog box will appear. From this dialog box, you can change Office Assistant options and choose a new Office Assistant personality.

1.5.6 Using the What's This Feature.

If you would like to learn about a button or other icon on the screen, then the *What's This* feature may be helpful. To use the *What's This* feature,

Step 1: Choose **Help** → **What's This** from the Menu bar. Your cursor should change shape to the *What's This* icon, ?.

Step 2: Move the mouse to the icon or graphic item that you want to learn about and click the left mouse button.

Step 3: A small help box will appear explaining the item, and your cursor will return to its original shape.

1.5.7 Learning from the Tip of the Day

If you would like Excel to provide you with a helpful tip each time you start up the Excel program, do the following:

Step 1: Open the Office Assistant.

Step 2: Choose **Options**. The Office Assistant dialog box will appear.

Step 3: Select the **Options** tab.

Step 4: Check the item labeled *Show the Tip of the Day at startup*.

1.5.8 Accessing Help from the World Wide Web

A wealth of information about Microsoft Excel is available on the World Wide Web. To access an Excel-related Web site,

Step 1: Make sure that you are connected to the Internet via a network or dial-up connection.

Step 2: Choose **Help** → **Office on the Web**.

You will be directed to the Microsoft Office Web site.

1.6 MANIPULATING WINDOWS

It is a good idea to spend some time getting used to manipulating windows before we actually begin to create a workbook. Excel allows you to keep more than one workbook open at a time and to have multiple worksheets within a single workbook. This is helpful when you are copying text or objects from one worksheet to another. However, having many windows open at one time can be confusing.

There are three important buttons on the Title bar of every workbook and the same three buttons appear on the main Excel Title bar. These buttons, called *Control buttons*, are used to control the window and are shown in Table 1.1. The functions of each Control button are explained next.

TABLE 1.1 The Window Control buttons

Button	Name
	Minimize Control button
	Restore/Maximize Control button
	Close Control button

1.6.1 Minimize Control Button

The *Minimize Control button* reduces the workbook to an icon near the bottom of the Excel document window. The small icon will resemble Figure 1.21, but the name of the workbook may be different. To restore the minimized workbook, double-click on the workbook's icon.

The minimized workbook icon may or may not show the Control buttons, depending on whether the icon is selected. Figure 1.21 shows two icons, one with the Control buttons and one without.

When minimized, a workbook is not closed, and it remains in computer memory. This is useful if you plan to use the document again shortly, since the document will not have to be retrieved from disk.

1.6.2 Restore/Maximize

The *Maximize Control button* allows you to maximize the physical area that your workbook or application occupies on the screen. Choose the Maximize Control button on the title bar of a window to expand the workbook or application to its maximum size. Choose the button again (now called the *Restore Control button*) to restore a full screen workbook or application to a smaller window size. Note that the Maximize Control button has two shapes and names, which alternate when a workbook is expanded or contracted. (See Table 1.1.)

1.6.3 Close Control Button

The *Close Control button* will close your workbook or application. If there are unsaved changes, then an Alert dialog box will appear and remind you to save your work. The Alert dialog box is shown in Figure 1.22. If you choose **No**, then the workbook or application will be closed without saving. If you choose **Cancel**, then the workbook or application will not be closed, and you will be returned to your application. If you choose **Yes**, the Save As dialog box will appear. We will discuss how to save your workbook in the next section.

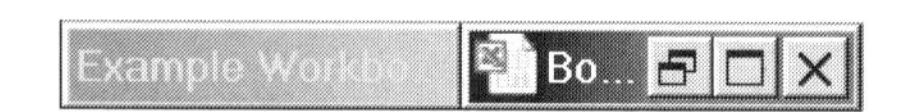

Figure 1.21. Example of two minimized workbooks.

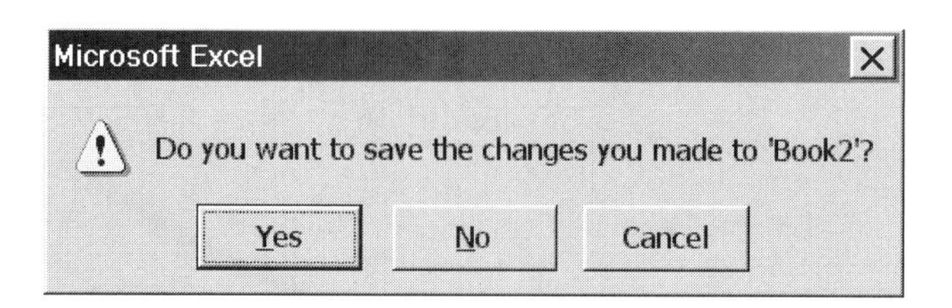

Figure 1.22. The Alert dialog box.

1.7 CREATING AND SAVING WORKSHEETS AND WORKBOOKS

1.7.1 Creating a New Workbook

When the Excel application is started, a blank workbook is automatically created. To create another new workbook,

Step 1: Choose **File** → **New** from the Menu bar.

Step 2: The Task pane will appear. Choose **New** → **Blank Workbook**.

Step 3: The *Task pane*, which allows you to have easy access to frequently used commands, will disappear and a new blank workbook will appear.

An alternate, quicker method for creating a new blank workbook is to choose the **New** icon on the Standard toolbar.

1.7.2 Opening an Existing Workbook

To open an existing workbook, do the following:

Step 1: Choose **File** → **Open** from the Menu bar. An alternative method is to choose the **Open** icon on the Standard toolbar.

Step 2: The Open dialog box will appear, as shown in Figure 1.23.

Step 3: Choose a file to open.

From the Open dialog box, you can type in a path and file name, or you can browse the file system to locate a file. The icons along the left side of the Open dialog box are used to help you find files quickly. By clicking on the icon labeled *History*, you will be shown the locations of your most recently used files. By clicking on the icon labeled *My Documents*, you will be taken to a special folder named *My Documents*. If you are working in a computer lab, be aware that the My Documents folder may be shared by other students. Ask your instructor where you should store your workbooks.

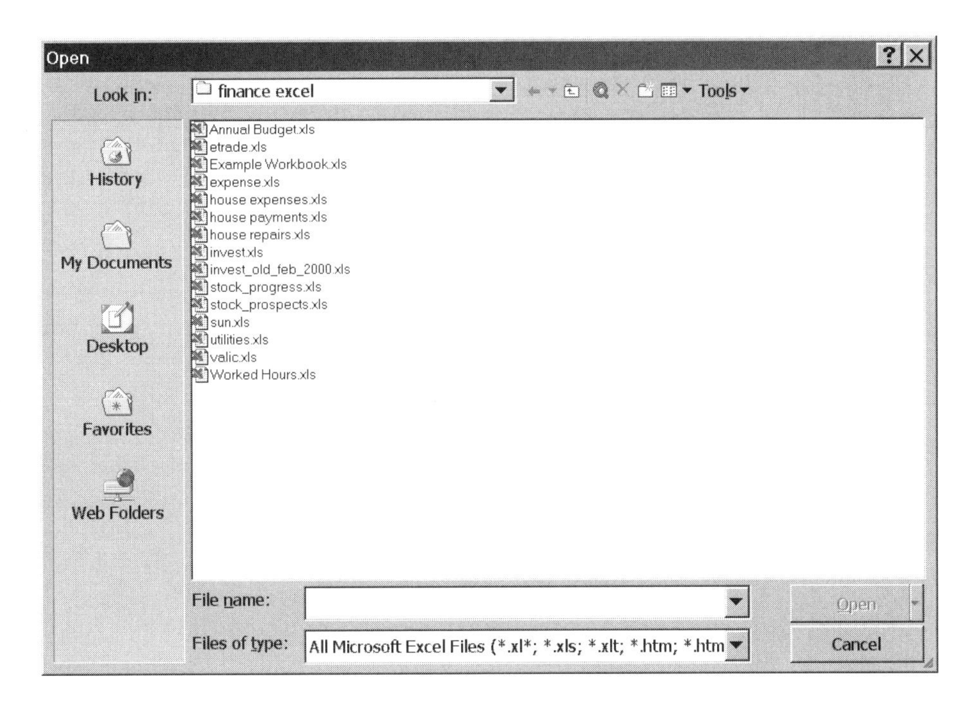

Figure 1.23. The Open dialog box.

1.7.3 Creating a New Worksheet

Within a workbook, you can have many worksheets. Usually, the worksheets in a workbook are all part of the same workbook file. However, you may link parts of worksheets to other files. The number of worksheets that you can have in a single workbook is limited only by the available memory on your computer.

To create a new worksheet in an open workbook, choose **Insert → Worksheet** from the Menu bar. An example of how you might use multiple worksheets is the creation of a lab report. Sheet 1 might be labeled *Lab Data*, sheet 2 labeled *Report*, sheet 3 labeled *Chart 1*, etc.

1.7.4 Introduction to Templates

A *template* is a workbook that has some of its cells filled in. If you use similar formatting for many documents, then you will benefit from creating and using a template. You may build your own template or customize preformatted templates and, in time, create a library of your own template styles. To open an example template,

Step 1: Choose **File → New** from the Menu bar.

The Task pane will appear.

Step 2: From the Task pane, choose **New from template → Loan Amortization**. A preassembled worksheet will appear. Fill in the blank cells labeled *Loan Amount, Annual Interest Rate*, etc. The worksheet will build an amortization table for you.

1.7.5 Opening Workbooks with Macros

A *macro* is a short computer program that records a group of tasks, which in turn is stored in a Visual Basic module. A set of frequently repeated commands can be stored and then executed with a single mouse click whenever needed.

Macros are very powerful tools. However, macros can contain viruses that will infect files on your computer. For this reason, you should only enable macros if you are certain of the origin of the macro. For example, if you followed the previous example and opened the Loan Amortization template, you were activating a macro. Since this template came with Excel, you can assume that it is safe. If you are unsure of the source of a macro, you should check the document by using a virus-protection software before opening the document. Virus-protection software is not provided with Microsoft Excel and must be purchased separately.

To change the macro security settings,

Step 1: Choose **Tools → Options** from the Menu bar.

Step 2: Click the **Macro Security** button. The Security dialog box will appear, as shown in Figure 1.24.

Step 3: Choose the **Security Level** tab.

Step 4: Check the box that matches your level of risk. It is recommended that you use at least *Medium*-level security.

1.7.6 Saving Documents

To save a document for the first time,

Step 1: Choose **File → Save As** from the Menu bar.

Step 2: The Save As dialog box will appear as shown in Figure 1.25.

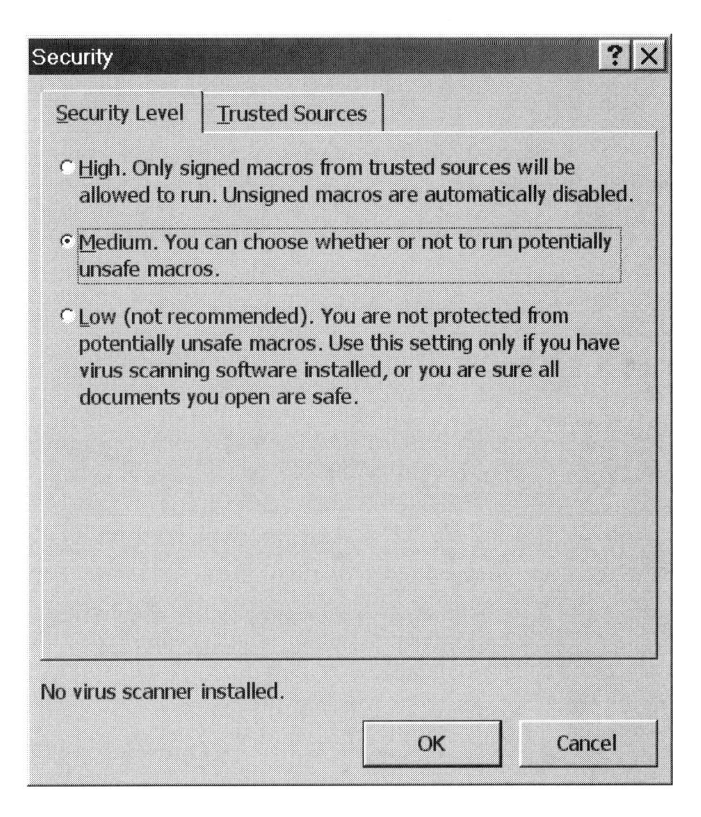

Figure 1.24. Setting the Macro Security Label.

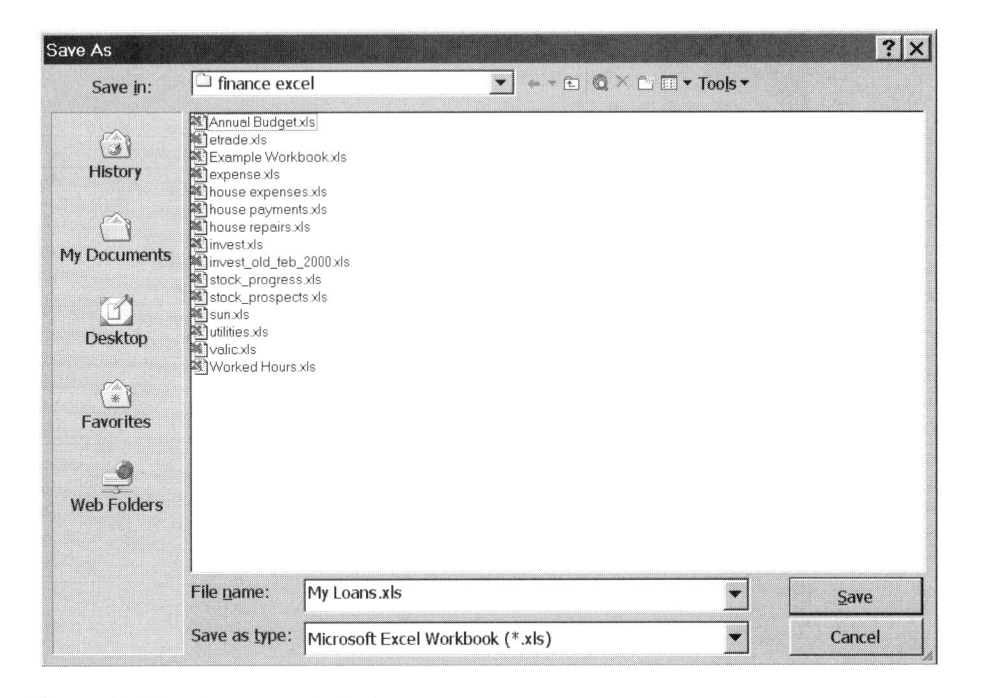

Figure 1.25. The Save As dialog box.

Step 3: Choose a folder in which to save the document by selecting the button on the right side of the box labeled *Save In*. Or you can choose one of the special folder icons along the left side of the dialog box.

Step 4: Type in (or select) a name for your document.

The document will be saved by default with the *XLS* file extension. Once a document has been given a name and saved, it may be reopened and edited.

To save an open document that already has a name, choose **File** → **Save** from the Menu bar or choose the **Save** icon from the Standard toolbar. If you are unsure of the name of the current working document, you can view it in the Title bar.

You should save your work frequently. It is also important to make backup copies of your important documents on floppy disks or some other physical device. There are many tales of woe from students (and professors) who have lost hours of work after a power failure.

1.7.7 The AutoRecover Feature

Fortunately, Excel has an automatic recovery feature, called *AutoRecover*, that makes the frequent saving of documents an easy task. However, the task of making frequent backup copies to a different medium (e.g., floppy disk), is something that you must perform manually.

To set the AutoRecover features,

Step 1: Choose **Tools** → **Options** from the Menu bar.

Step 2: Choose the **Save** tab. The Options dialog box will appear as shown in Figure 1.26.

Step 3: Check the box labeled *Save AutoRecover info every*.

Step 4: Choose a time interval for automatic saves.

Step 5: Choose a location for the AutoRecover information. It is wise for this location to be on a different physical disk from that where you store your documents.

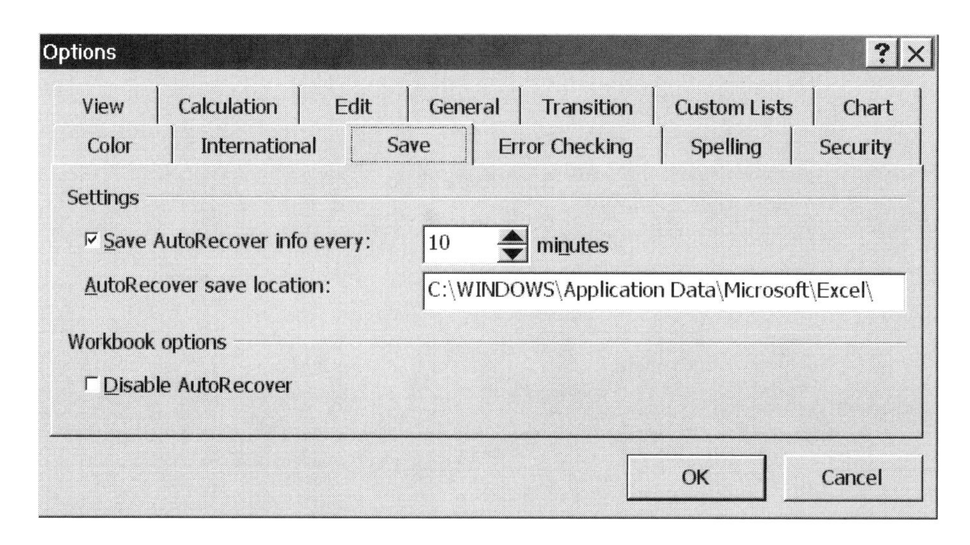

Figure 1.26. The Save tab of the Options dialog box.

While you have the Options dialog box open, take some time to view the other user options that may be customized. Browse through the other tabs on the Options dialog box. Until you become more familiar with Excel, you should probably leave most of the options set to their default values.

1.7.8 Naming Documents

It is important to develop a methodical and consistent method for naming worksheets. Over time, the number of worksheets that you maintain will grow larger, and it will become harder to locate or keep track of them. Documents that are related should be grouped together in a separate folder. Do not use the default workbook names (i.e., Book1, Book2, Book3, etc), or chaos will soon ensue.

If documents are not given meaningful names, then the documents may be inadvertently overwritten. Documents that have very general names (e.g., *Spreadsheet*), will be difficult to locate later.

By default, Excel documents are given the extension *XLS*. Unless you are specifically creating a template (*XLT*), ASCII text document (*TXT*), or other special type of document, you should use the default extension.

1.8 MOVING AROUND A WORKSHEET

There are several methods of moving from place to place in an Excel worksheet. If the worksheet is relatively small, all of these methods will work equally well. As a worksheet grows in size, movement becomes more difficult, and you can save a lot of time by learning the various movement methods.

The current cell number is displayed in the Name box on the left-hand side of the Formula bar. Place the cursor over column H and row 5 of a new worksheet. The currently active cell location will appear in the Name box. This is demonstrated in Figure 1.27.

The three general methods for moving around a document are as follows:

- Movement by using the keyboard
- Movement by using the mouse
- Movement by using the Go To dialog box

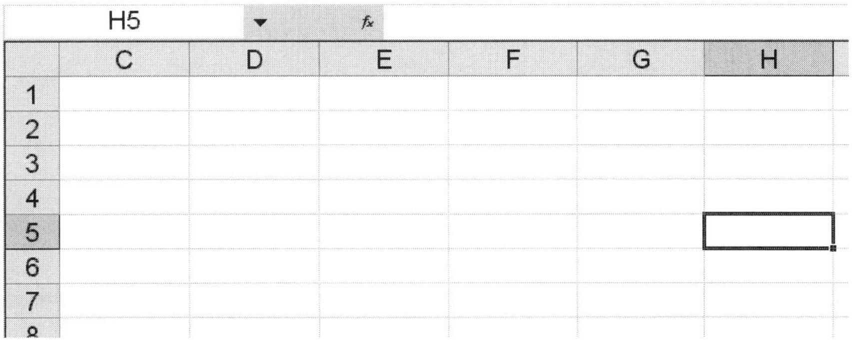

Figure 1.27. The Name box.

1.8.1 Movement by Using the Keyboard

The keyboard may be used to select a worksheet from a workbook. The keyboard may also be used to navigate around a single worksheet quickly and effectively. You may already use the arrow keys to move up, down, left, and right. Combining the **Ctrl** key with the arrow keys gives you the means for rapid movement. Table 1.2 lists the most frequently used key combinations for movement.

PRACTICE!

Step 1: Open a new workbook.
Step 2: Create several worksheets in the workbook by choosing **Insert → Worksheet** from the Menu bar.
Step 3: Practice each of the keyboard movement commands in Table 1.2.
Step 4: Move to the far-right column and bottom row of a worksheet. What is the maximum size of a worksheet?

1.8.2 Movement by Using the Mouse

The mouse may be used to select a worksheet and to move within a worksheet. To select a worksheet, choose a tab from the Sheet Tab bar as depicted in Figure 1.14.

One method of moving around a worksheet with the mouse is to click on a cell. This is most useful if the new insertion point is located on the same screen. If the desired location is on a different page, then the Vertical and Horizontal scrollbars may be used to move quickly to a distant location. Figure 1.15 shows the Vertical and Horizontal scrollbars.

1.8.3 Movement by Using the Go To Dialog Box

If you have a large worksheet that covers many screens, then using the keyboard and mouse can be a cumbersome way of moving through the worksheet. The Go To dialog box offers a method for moving directly to distant locations on the worksheets.

To move to a location using the Go To feature,

Step 1: Open the Go To dialog box by choosing **Edit → Go To** from the Menu bar. Alternatively, you can press the **F5** key. The Go To dialog box will appear, as depicted in Figure 1.28.

Step 2: Type in a cell reference. For example, type **H5**, then click **OK**. The screen will move to cell H5, and it will become the active cell.

TABLE 1.2 Movement within a Worksheet, Using the Keyboard

Key Combination	Action
←	Move one cell (column) to the left
→	Move one cell (column) to the right
↑	Move up one cell (row)
↓	Move down one cell (row)
Ctrl + →	Move to far-right column of worksheet
Ctrl + ↓	Move to bottom row of worksheet
Page Down	Move down one screen
Page Up	Move up one screen
Ctrl + **Page Down**	Select next worksheet
Ctrl + **Page Up**	Select previous worksheet

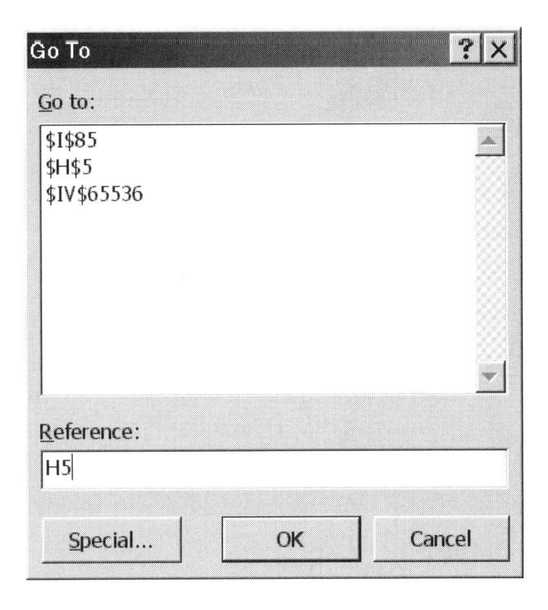

Figure 1.28. The Go To dialog box.

A history of previous references is kept in the Go To window, so recently visited cells can be quickly located simply by selecting them with the mouse.

In addition to moving to cells by location, you can move to cells of a particular type. We have not yet shown you how to create cells of different types. However, imagine that you have created a number of cells containing formulas. You can locate formulas with errors in them by using the Go To Special dialog box as follows:

Step 1: Click the **Special** button on the Go To dialog box. The Go To Special dialog box will appear, as shown in Figure 1.29.

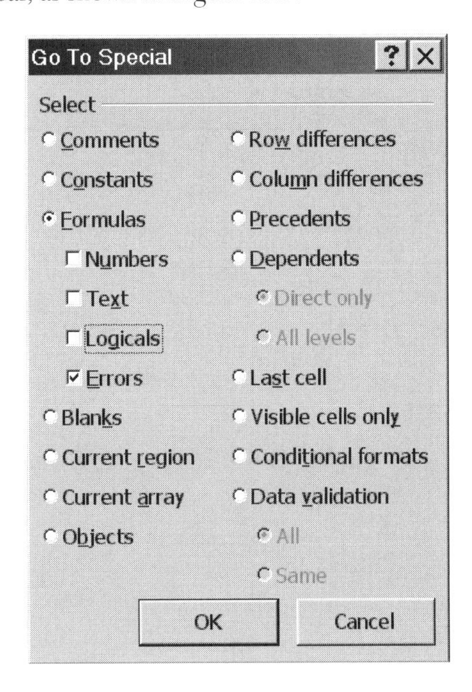

Figure 1.29. The Go To Special dialog box.

Step 2: Check the box labeled *Formula* and the box labeled *Errors*.

Step 3: Uncheck all other boxes and click **OK**.

The first formula with an error will become the active cell, and all other formulas with errors will be highlighted.

1.9 SELECTING A REGION

Much of the time spent in worksheet preparation involves moving, copying, and deleting regions of cells or other objects. In this section, we will be selecting regions of cells, but the same principles apply to regions that contain charts, formulas, and other objects. Before an action can be applied to a region, the region must be selected. The selection process can be performed by using either the mouse or the keyboard.

1.9.1 Selection by Using Cell References

In many cases, you will have the option of typing a cell reference. For example, you can type cell references into a formula. A single cell is noted by its column letter and row number. A rectangular range of cells is denoted by the reference for the top-left and bottom-right cells. For example, the rectangle bordered by A1 on the top left and D5 on the bottom right is denoted as A1:D5.

1.9.2 Selection by Using the Mouse

To select a region of text with the mouse, move the mouse to the beginning of the region, click it, and drag it to the end of the region. As you drag the mouse, the selected region will be highlighted.

To select a region of text that is larger than one screen, drag the mouse to the bottom of the screen. If you hold the mouse at the bottom of the screen without releasing the mouse button, then the selected region will continue to grow and the screen will scroll downward. This takes a little practice.

To select a whole column, click on the column header. To select a whole row, click on the row header.

To select the entire worksheet, choose the button at the top-left corner of the worksheet. This blank button is called the Select All button. The select-all button is depicted in Figure 1.30. This is useful if you are applying a change to every cell in a worksheet.

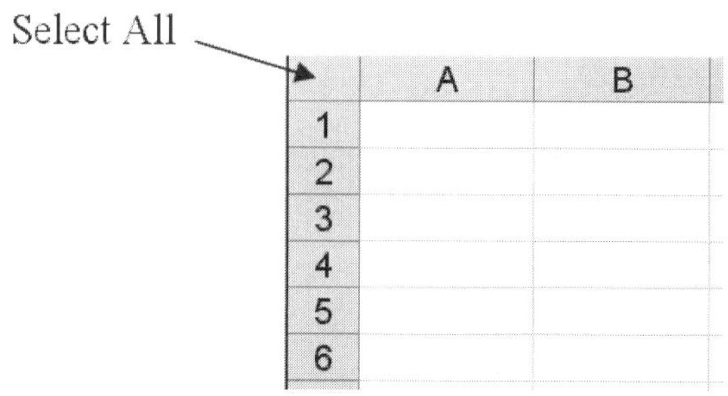

Select All

	A	B
1		
2		
3		
4		
5		
6		

Figure 1.30. The Select All button.

1.9.3 Selection by Using the Keyboard

An alternate method for selecting regions of a document is to use the keyboard, as follows:

Step 1: Click the mouse on one corner of the region that you wish to select.

Step 2: While holding down the **Shift** key, use one of the movement keys shown in Table 1.2 to move to the other end of the region.

Step 3: The selected region will be highlighted.

Step 4: If you make a mistake and incorrectly select a region, then click the mouse cursor anywhere on the document window before you apply an action (such as delete). If the highlighting disappears, then you have deselected the region.

PRACTICE!

Try the following exercise to practice selecting regions:

Step 1: Place the cursor over cell B3 and type the number 5.

Step 2: Press the down-arrow key.

Step 3: Type the number 6.

Step 4: Press the down-arrow key.

Step 5: Type the number 7.

Step 6: With the mouse, place the cursor over cell B3, hold down the left-mouse button, and drag the mouse until cells B3, B4, and B5 are all highlighted.

Step 7: Choose the AutoSum icon Σ on the Standard toolbar.

A new cell B6 will be added that contains the sum of cells B3, B4, and B5. The results should resemble Figure 1.31.

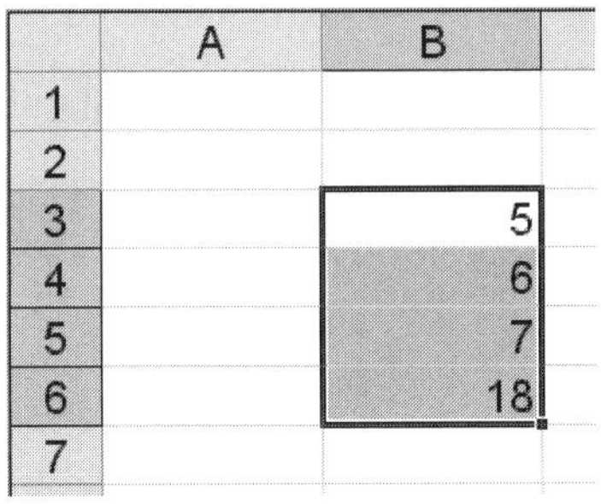

	A	B
1		
2		
3		5
4		6
5		7
6		18
7		

Figure 1.31. Example of region selection.

1.10 CUTTING, MOVING, COPYING, AND PASTING

Once a region has been selected, several actions may be taken, such as delete, move, copy, and paste. As usual, Excel provides several ways to accomplish the same actions. These include using keyboard commands and mouse commands.

The cut, copy, and paste commands make use of a special location called the *clipboard*. The clipboard is a temporary storage location that can be used to hold a region of cells, or most other objects, such as pictures. To view the contents of the clipboard, make sure that your Task pane is turned on. To turn on the clipboard section of the Task pane, perform the following steps:

Step 1: Choose **View** from the Menu bar.

Step 2: Make sure that the item labeled *Task Pane* is checked.

Step 3: Select the small down arrow in the top-right corner of the Task pane. A drop-down menu will appear.

Step 4: Check the item labeled *Clipboard*.

1.10.1 Cutting a Region

Cutting a region places it on the clipboard. A region may be cut using the mouse or the keyboard. To cut a region by using the mouse,

Step 1: Select a region.

Step 2: Choose **Edit** → **Cut** from the Menu bar. The cut region will be highlighted by a rotating dashed line. An alternate method is to click the right mouse button and then choose **Cut** from the Quick Edit menu that appears. The Quick Edit menu is shown in Figure 1.32. Yet, a third alternate method is to choose the **Cut** icon from the Standard toolbar.

To cut a region using the keyboard,

Step 1: Select the region.

Step 2: Press **Ctrl** + **X.**

No matter which method you use to cut the region, the effect is to place the contents of the region on the clipboard. This will be displayed in the Task pane. If you cut the region used in the previous **PRACTICE** box, the results will appear, as shown in Figure 1.33.

1.10.2 Moving a Region

A region may be moved by first cutting the region and then *pasting* it to a new location. To paste a selection by using the mouse,

Step 1: Select and cut a region.

Step 2: Select a destination cell or region.

Step 3: Click the right mouse button, and choose **Paste** from the Quick Edit Menu. The original region of cells should now appear in the new location. If you do not create the new region with the same size and shape, then Excel will create a region of appropriate size. An alternate choice for this step is to choose the **Paste** icon from the Standard toolbar.

By using the cut and paste method of moving a region, you can move the region across documents and even across applications. For example, you can cut a region from an Excel worksheet and paste it into a Word document.

To paste a region by using the keyboard,

Step 1: Select a region.

Step 2: Press **Ctrl** + **X** to cut the region to the clipboard.

Step 3: Select a destination cell or region.

Step 4: Press **Ctrl** + **V** to paste the region.

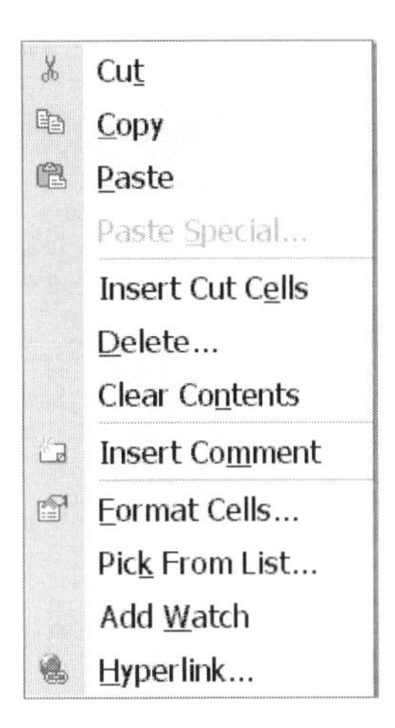

Figure 1.32. The Quick Edit menu.

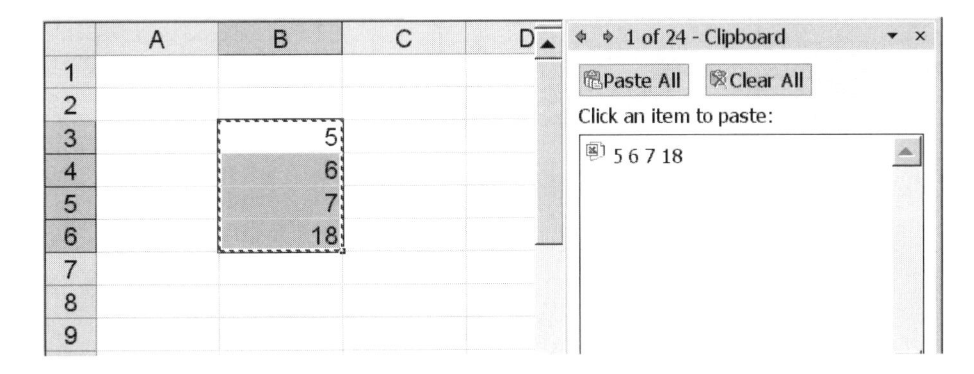

Figure 1.33. Cutting a region and placing it on the clipboard.

1.10.3 Copying a Region

Copying a region is similar to moving a region, except that the original copy of the region remains intact. To copy a region by using the mouse;

Step 1: Select the region to be copied.

Step 2: Choose **Edit** → **Copy** from the Menu bar. Or click the right mouse button and choose **Copy** from the Quick Edit menu.

Step 3: Select a destination cell or region.

Step 4: Choose **Edit** → **Paste** from the Menu bar to make a copy. Or choose **Copy** from the Quick Edit menu. The last action can be performed as many times as needed if multiple copies are to be made.

The copy and paste method can be used to copy regions to another document or another application. To copy a region by using the keyboard,

Step 1: Select a region.

Step 2: Press **Ctrl** + **C** to copy the region to the clipboard.

Step 3: Select a destination cell or region.

Step 4: Press **Ctrl** + **V** to paste the copy.

1.11 INSERTING AND DELETING CELLS

New cells may be added to a worksheet, and cells may be deleted. To delete a region of cells,

Step 1: Select the region.

Step 2: Choose **Edit** → **Delete** from the Menu bar. Or choose **Delete** from the Quick Edit menu. In either case, the Delete dialog box will appear, as shown in Figure 1.34.

Step 3: Choose whether you want to delete an entire row or column, or whether you want the remaining cells to be shifted left or up after the deletion.

To clear the contents of a region of cells without shifting the cells, choose **Edit** → **Clear** → **Contents** from the Menu bar. Or choose **Clear Contents** from the Quick Edit menu.

You can insert new cells, rows, columns, or an entire worksheet by selecting **Insert** from the Menu bar.

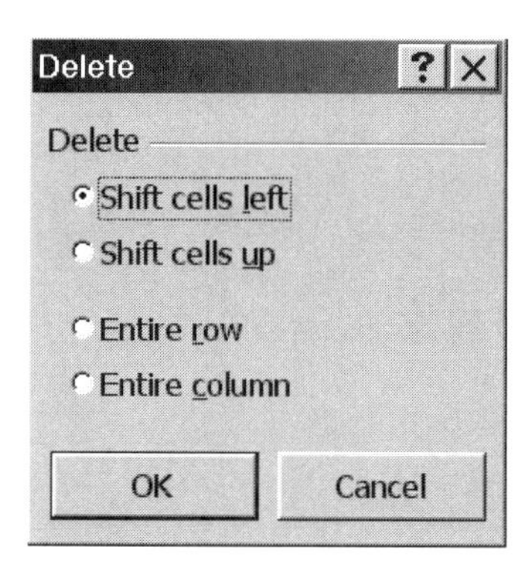

Figure 1.34. The Delete dialog box.

1.12 SHORTCUT KEYS

As a novice user, you may have trouble finding commands. Using the Menu bar is a good method in the beginning, since the command names are listed and the commands are usually in the same place. As you gain more experience, the toolbars become more useful, because a toolbar button is faster to execute than a Menu bar selection. As you become very proficient with Excel and learn to type at a rapid rate, the *shortcut keys* become the quickest way to execute a command. Movement of the fingers from the keyboard to the mouse is avoided. The downside of using keyboard shortcuts is that they have to be memorized.

We have shown you several shortcut key combinations in the preceding sections. As you progress with Excel, learning a few of the most commonly used shortcut keys can save you a lot of time.

One method of learning some of the shortcuts is to look at the right-hand side of the Menu bar items. For example, choose **Edit** from the Menu bar and note that the shortcut for *Cut* is listed on the menu as *Ctrl* + *X*. Table 1.3 shows some of the most commonly used keyboard shortcuts.

TABLE 1.3 Commonly Used Shortcut Keys

Command	Shortcut
File → New Workbook	Ctrl + N
File → Open Workbook	Ctrl + O
File → Save Workbook	Ctrl + S
File → Print	Ctrl + P
Edit → Undo	Ctrl + Z
Edit → Copy	Ctrl + C
Edit → Paste	Ctrl + V
Edit → Find	Ctrl + F
Edit → Replace	Ctrl + H
Edit → Go To	Ctrl + G
Format → Cells	Ctrl + 1
Help	F1
Find and Replace	F5
Tools → Spelling and Grammar	F7

1.13 FINDING AND CORRECTING MISTAKES

1.13.1 Undoing Mistakes

Excel allows actions to be undone or reversed. To undo the last action, choose **Edit → Undo** from the Menu bar or type **Ctrl + Z.**

A third method for undoing or redoing a command is to use the Undo or Redo icons on the Standard toolbar. The Undo and Redo icons are shown in Figure 1.35.

To see the list of recent actions, choose the down-arrow button next to the Undo icon. From this list, you may select one or more actions to be undone. Note that if you select an action on the list, then all of the actions preceding it will also be undone! If you accidentally undo an action, then you may redo it by selecting the redo button.

1.13.2 Checking Spelling

Excel can check the spelling of cells containing text. To check the spelling in a region, first select the region, then choose **Tools → Spelling**. Alternate methods are to press the **F7** key or choose the **Spelling** icon from the Standard toolbar. If Excel finds a spelling mistake, then the Spelling dialog box will appear, as shown in Figure 1.36.

The region containing the mistake is displayed in the top text box. Suggestions for changes are presented in the bottom text box. At any point in the process, you can choose whether to accept or ignore the suggestions. If you choose a suggested correction, then you may click the **Change All** button to change all occurrences of the misspelled word in the selected region.

You may add new words to the main dictionary by choosing the **Add** button. This will probably be necessary as you proceed through your coursework, since many engineering terms are not in the custom dictionary.

1.13.3 The AutoCorrect Feature

The Excel *AutoCorrect* feature recognizes spelling errors and corrects them automatically. AutoCorrect performs actions such as automatically capitalizing the first letter of a sentence or correcting a word whose first two letters are capitalized.

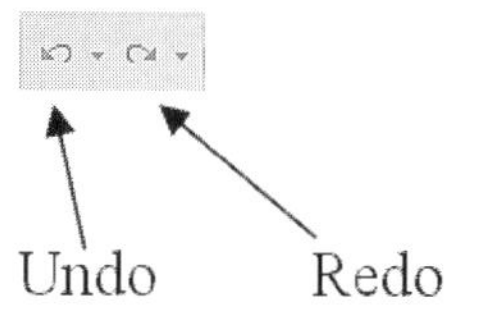

Figure 1.35. The Undo and Redo icons from the Standard toolbar.

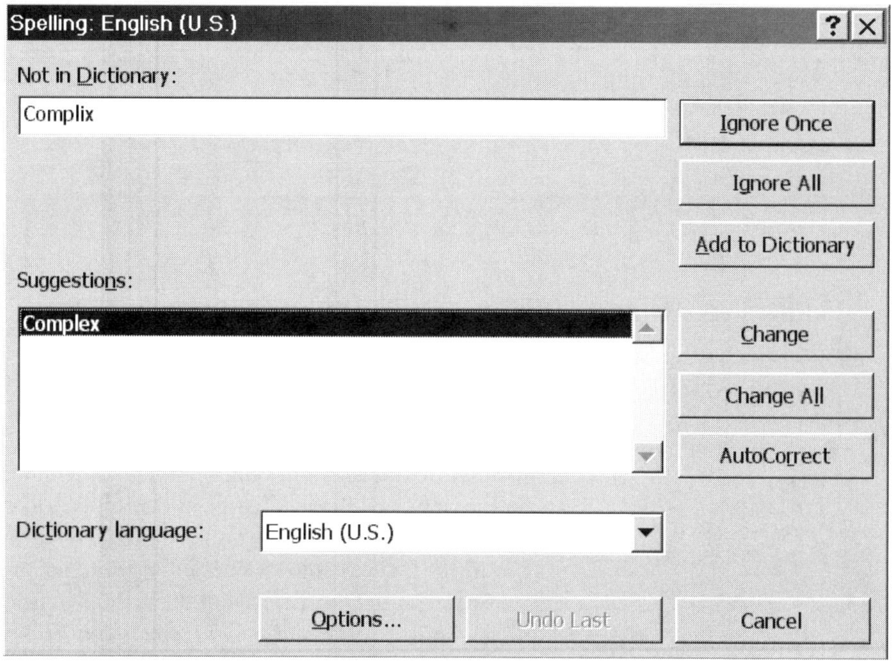

Figure 1.36. The Spelling dialog box.

You can test to see if the AutoCorrect feature is turned on for your installation of Excel. Try typing the letters *yuo*, then press the spacebar. Was the word automatically retyped as *you*? If so, then you have AutoCorrect turned on.

To see your AutoCorrect settings and dictionary, choose **Tools → AutoCorrect**. The AutoCorrect dialog box will appear, as shown in Figure 1.37.

From the AutoCorrect dialog box, you can select (or deselect) various options. You can also scroll through the AutoCorrect dictionary, add entries to the dictionary, and add exceptions to the dictionary. Creating an exception list will be necessary if you use all of the AutoCorrect features. For example, if you have selected the option that automatically converts the second capital letter to lowercase, you may have an occasional exception. Be careful when adding new entries into the AutoCorrect dictionary. You may inadvertently add an entry for a misspelling that is a legitimate word.

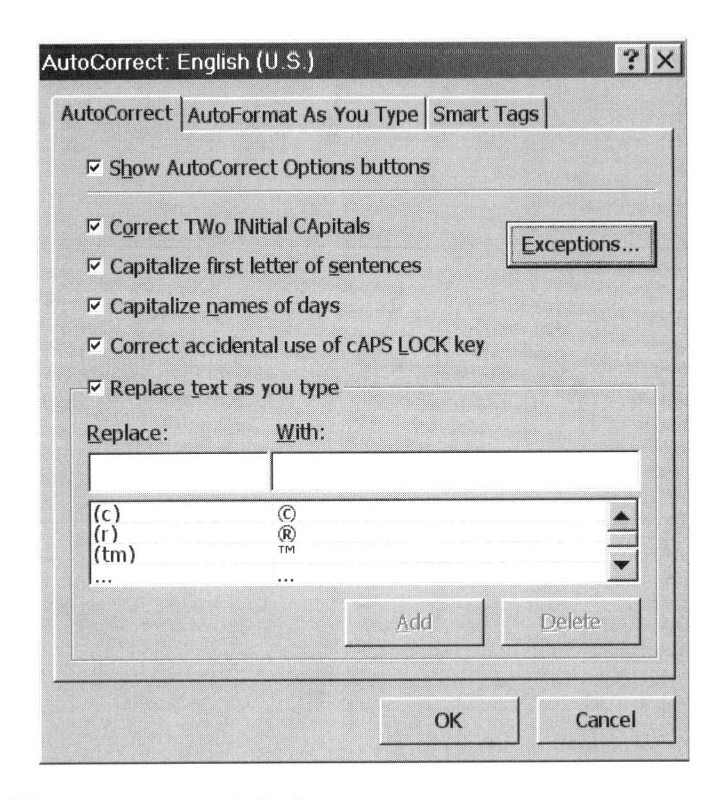

Figure 1.37. The AutoCorrect dialog box.

1.14 PRINTING

Before attempting to print a document, make sure that your printer is correctly configured. See your operating system and printer documentation for assistance.

1.14.1 Setting the Print Area

If you want to print a region of a worksheet instead of the whole worksheet, you must first set the print area. To set the print area,

Step 1: Select a region to print.

Step 2: Choose **File** → **Print Area** → **Set Print Area** from the Menu bar. The selected region will now be surrounded by a dashed line.

1.14.2 Previewing a Worksheet

It is advisable to preview a document before printing it. Many formatting problems can be resolved during the preview process. To preview the document as it will be printed,

> Select **File** → **Print Preview** from the Menu bar. Or choose the **Print Preview** icon from the Standard toolbar. The selected region will be displayed in the same format in which it will be printed. In addition, the Print Preview menu bar is placed on the screen.

The Print Preview menu bar is displayed in Figure 1.38. The available options on the Print Preview menu bar are summarized in Table 1.4.

Figure 1.38. The Print Preview menu bar.

TABLE 1.4 Print Preview Menu Bar Options

Button	Action
Next	Display next page of the worksheet
Previous	Display previous page of the worksheet
Zoom	Toggle between magnified and normal view
Print	Print the worksheet
Setup	Set the page orientation, margins, page order, etc.
Margins	graphically set the margins and page stops
Page Break Preview	Graphically set the page breaks
Close	Close this window and return to the worksheet
Help	Special help for the Print Preview menu bar

1.14.3 Printing a Worksheet

You can print a worksheet in at least four different ways. To print a worksheet,

Step 1: Choose **File** → **Print** from the Menu bar. The Print dialog box will appear as depicted in Figure 1.39.

Step 2: Or select the **Print** icon from the Standard toolbar. The job will be sent directly to the printer without going through the Print dialog box.

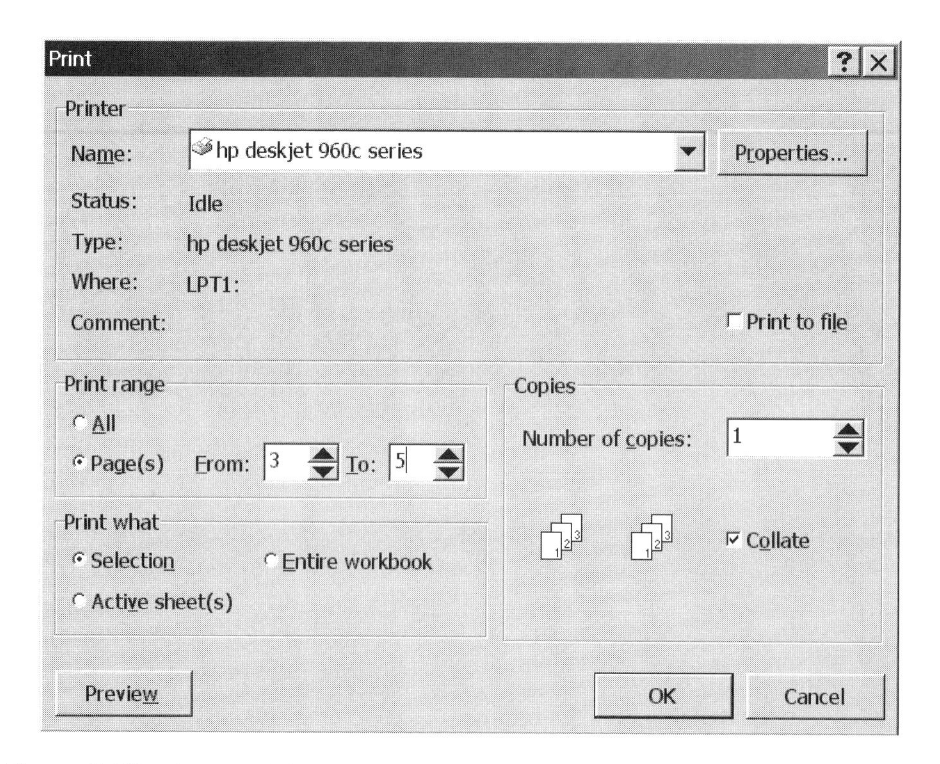

Figure 1.39. The Print dialog box.

Step 3: Or choose **Print** from the Print Preview menu bar. The Print dialog box will appear.

Step 4: Or press **Ctrl + P**. The Print dialog box will appear.

The Print dialog box includes commands for collating, selecting the number of copies to print, and selecting a range of pages. The Properties button will yield in a set options that depends on the type of printer that you have connected to your computer.

If you have made some modifications on a number of pages of your workbook, you can save a great deal of paper by using the Print Preview feature and then selecting and using the option labeled *Print range* on the Print dialog box.

KEY TERMS

AutoCorrect	AutoRecover	cells
clipboard	Close Control button	Control button
cutting	Formula bar	macro
Maximize Control button	Menu bar	Minimize Control button
pasting	scrollbars	shortcut key
spreadsheet	Status bar	Task pane
template	Title bar	Tool tip
toolbars	workbook	worksheet

SUMMARY

In this chapter, you were introduced to Microsoft Excel. The basic Excel components, including the Title bar, Menu bar, scroll bars, and various toolbars, were presented. You were shown methods for accessing on-line help. You were guided through the creation of a new worksheet and the basic commands for editing and printing a worksheet.

Problems

1. Test your understanding by filling in the blanks.
   The _____ _____ displays the name of the currently open workbook. The View, Insert, and Format items are found on the _____ _____. Clicking on the Open icon on the Standard toolbar has the same effect as choosing _____ and then _____ from the Menu bar. You can add or delete toolbars by choosing _____ and then _____ from the Menu bar.
2. What is the maximum number of rows and columns for a single Excel worksheet?
3. Use the Insert Function dialog box to identify the Excel function names for the following mathematical functions:

   _____ sine
   _____ arithmetic mean
   _____ natural logarithm
   _____ convert degrees to radians
   _____ remove or truncate the decimal part of a number
   _____ return e raised to the power of a number

4. Name three ways to undo a mistake.
   _____, _____, _____

5. Name the shortcut keys for the following actions:

 _____ Help
 _____ Copy selected region
 _____ Cut select region
 _____ Move to the beginning of a worksheet (see Help for this one)

6. Visit the U.S. National Institute of Science and Technology (NIST) Physics Laboratory's Web site about the International System of Units (SI) at

 http://physics.nist.gov/Divisions/Div840/SI.html

 Click on the menu item labeled *In-depth Information on the SI, the Modern Metric System*, and locate the table for SI Base Units. Use that table to fill in the missing entries in Table 1.5.

TABLE 1.5 SI Base Units

Quantity	Name	Symbo
length		m
	kilogram	kg
time	second	
electric current	ampere	
temperature		K
	mole	mol
luminous intensity		cd

7. The electronic spreadsheet has played an important role in the history of computing. The links presented here discuss the history of electronic spreadsheets. Access these Web sites with your Web browser and then answer the questions that follow:

 Power, Daniel. *A Brief History of Spreadsheets*, at URL http://www.dssresources.com/ history/sshistory.html.

 Browne, Christopher. *Historical Background on Spreadsheets*, at URL http://www.ntlug.org/~cbbrowne/spreadsheets.html.

 Mattessich, Richard. *Early History of the Spreadsheet*, at URL http://www.j-walk.com/ss/history/spreadsh.htm.

 - What is the name of the first marketed electronic spreadsheet that was partly responsible for the early success of the Apple computer?
 - What spreadsheet application remains the most widely sold software application in the world (as of 1998)?

8. Excel's trigonometric function PI returns an approximation of the mathematical constant π. Read the bottom of the Insert Function dialog box to determine the number of digits of accuracy of the constant returned by this function.

2

Entering and Formatting Data

SECTION

OBJECTIVES

After reading this chapter, you should be able to:

- Enter numeric symbols, text, and time and date data into a worksheet.
- Quickly enter series of data.
- Format rows, columns, and cells.
- Apply conditional formatting to a range of cells.
- Sort one or more columns of cells.
- Apply formatting to an entire worksheet.

2.1 ENTERING NUMERIC DATA

Worksheet cells can be filled with numeric values, text, times, dates, logical values, and formulas. In addition, a cell may contain an error value if Excel cannot evaluate its contents. The most common type of data entered into a spreadsheet is numeric data.

Numeric values containing any of the following symbols can be entered into a cell:

0 1 2 3 4 5 6 7 8 9
+ − ( ) , /
$ % .
E e

When formatting your cell, you give Excel the capability to change the way that numbers and text are displayed in a worksheet. A format changes only the appearance of a number, not its value. Here's a list and the definition of the most common type of formatting commands you may use to properly represent your data in Excel:

- *General format* means that Excel will choose a format base on the contents of the cell.
- *Currency format* means that Excel will give the numeric data a monetary value.
- *Date format* means that Excel will display the data as a date.
- *Scientific format* means that Excel will display the number in scientific notation.
- *Text format* means that the contents of the cell will be treated as text.
- *Time format* means that the contents of the cell will be treated as time.

Numeric values are stored internally with up to 15 digits of precision (including the decimal point). As explained earlier, regardless of the internal storage, a numeric value can be displayed in a variety of formats. To see a list of cell formats,

Step 1: Type a number into a cell and then click the right mouse button. The Quick Edit menu will appear.

Step 2: Select **Format Cells** from the Quick Edit menu. The Format Cells dialog box will appear, as shown in Figure 2.1.

Step 3: Choose the **Number** tab.

From the Number tab of the Format Cells dialog box, the numeric value that you entered may be formatted in a variety of ways, e.g., currency, as a date, as a fraction, or in scientific notation. When you change the format, the internal representation is not changed; only the method for displaying the value is modified.

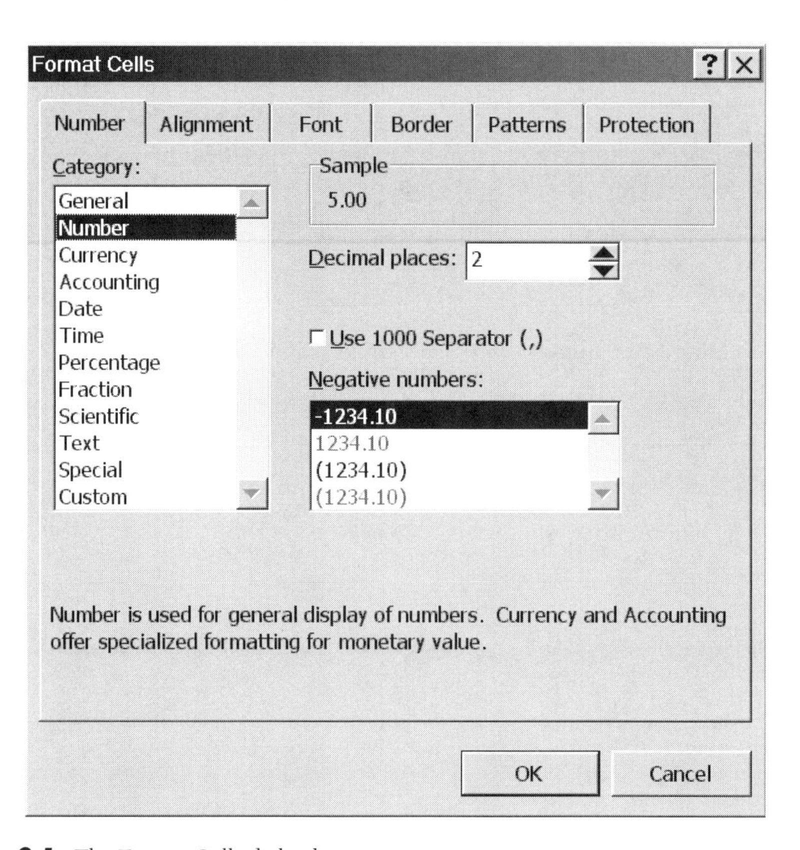

Figure 2.1. The Format Cells dialog box.

If the General format is selected, Excel will attempt to guess a format based on the contents of the cell. For example, if you type $3.4 into a cell, Excel will assume that you are entering a monetary value and will automatically convert to Currency format, and the value will be right justified and displayed as $3.40.

PRACTICE!

Open a blank worksheet and practice formatting a numeric value.

Step 1: Enter **12023.45** into cell B1.

Step 2: Type the same value into cells B2 : B4.

Step 3: By default, the cells are formatted in General format.

First, format cell B2 in Number format.

Step 4: Select cell **B2**, and choose **Format** → **Cells** from the Menu bar. The Format Cells dialog box will appear. Choose the **Number** tab and select **Number** from the list labeled *Category*.

Step 5: Set the scroll box titled *Decimal places* to **0**, and check the box labeled *Use 1000 Separator*. Note that this action rounds off the number to zero decimal places. However, no information is lost internally. If you were to change the number of decimal places to 2, the fractional part of the number would be displayed.

Next, format cell **B3** in Currency format.

Step 6: Select cell **B3**, and choose **Format** → **Cells** from the Menu bar. The Format Cells dialog box will appear. Choose the **Number** tab and select **Currency** from the list.

Step 7: Select **2** decimal places and select the **$** symbol.

Next, format cell **B4** in Scientific format. This will display the number in scientific notation, base 10. For example, the number 256 is equal to 2.56×10^2, which is represented in Scientific format as 2.56E+02.

Step 8: Select cell **B4**, choose **Format** → **Cells** from the Menu bar. The Format Cells dialog box will appear. Choose the **Number** tab and select **Scientific** from the list.

Step 9: Select **6** decimal places.

Your resulting worksheet should resemble Figure 2.2.

	A	B
1		12023.45
2		12,023
3		$12,023.45
4		1.202345E+04

Figure 2.2. The number 12023.45 in various formats.

2.2 ENTERING TEXT DATA

To treat the contents of a cell or region as text,

Step 1: Select the cell or region to be formatted.

Step 2: Choose **Format** → **Cells** from the Menu bar. The Format Cells dialog box will appear. Choose the **Number** tab and select **Text** from the list.

Step 3: Enter text into the cell or cells.

By default, Text-formatted cells are left justified. If you make a typing mistake, double click on the cell, and you will be able to edit the text.

Numeric values that are entered into a text cell cannot be used for calculation. In fact, Excel treats a numeric value that is entered in *Text format* as an error. Usually, you will be entering numbers in numeric format. If you want to quickly enter a number in Text format, type a single apostrophe before the number.

You can apply many of the same formatting options to the contents of a text cell that are used for a Word document. These include changing font size, font type, and performing spelling checks. To change the appearance of a Text-formatted cell, choose the **Font**, **Alignment**, and **Border** tabs on the Format Cells dialog box.

PRACTICE!

Practice adding text data to the worksheet in Figure 2.2.

Step 1: Select the region A1: A4.

Step 2: Choose **Format** → **Cells** from the Menu bar. The Format Cells dialog box will appear. Choose the **Number** tab and select **Text** from the list labeled *Category*.

Step 3: Enter the text *General* into cell A1.

Step 4: Enter the text *Number* into cell A2.

Step 5: Enter the text *Currency* into cell A3.

Step 6: Enter the text *Scientific* into cell A4.

Your worksheet should resemble Figure 2.3.

	A	B
1	General	12023.45
2	Number	12,023
3	Currency	$12,023.45
4	Scientific	1.202345E+04

Figure 2.3. Example of Text-formatted cells.

2.3 ENTERING DATE AND TIME DATA

Excel stores dates and times internally as numbers. This allows you to perform arithmetic on dates and times. For example, you can subtract one date from another. If a dash (−) or slash (/) is inserted between two digits, then Excel assumes the number is a date. If a colon (:) is used to separate two digits, then Excel assumes that the number represents time. Excel also recognizes the key characters AM, PM, A, and P to represent A.M. and P.M.

To enter data in Time or Date format,

Step 1: Select the cell or region to be formatted.

Step 2: Choose **Format** → **Cells** from the Menu bar. The Format Cells dialog box will appear.

Step 3: Choose the **Number** tab and select **Time** or **Date** from the list labeled *Category*.

Step 4: Enter the time or date data.

A shortcut method is to let Excel try to figure out whether you are entering time or date data. For example, if you type 3/2/2002 into a cell, Excel will assume that it is a date and will convert the entry to Date format. If you type 8:23 into a cell, Excel will assume that you mean 8:23:00 AM.

This feature can be frustrating. If you intend to type the fraction 4/5 into a cell, Excel assumes that you mean April 5 of this year. If you then try to convert the cell's format to the Fraction format, Excel converts the internal time representation to a fraction resulting in 37351. To type a fraction into a cell, you must first select the formatting to be Fraction and then enter the data!

2.4 USING FILL HANDLES

Data entry can be tedious. The use of *Fill Handles* allows one to quickly copy a cell into a row or column of cells. Fill Handles can also be used to create a series of numbers in a row or column. The Fill Handle appears as a small black square in the bottom-right corner of a selected region. An example fill handle is shown in Figure 2.4.

When the mouse is placed over the Fill Handle, its shape will change to a black cross. Click and hold the right mouse button while dragging the mouse to the right over four or five cells. When the mouse is released, the value in the original cell will be copied into the new cells.

While the Fill Handle command is used to quickly copy the contents of one cell to an adjacent cell, row, or column, the *Fill Series* command is used to extend or increment the values (linearly or exponentially) of the adjacent cells, rows, or columns. To see the Fill Series options, place the mouse over a fill handle and then click the right mouse button and drag the mouse. When the mouse is released, the Fill Series drop-down menu will appear, as shown in Figure 2.5.

The Fill series options can be changed by choosing Series from the drop-down menu. The Series dialog box will appear as shown in Figure 2.6. Getting used to Fill Handles and Fill Series takes a little practice, because left clicks, right clicks, and dragging must be coordinated.

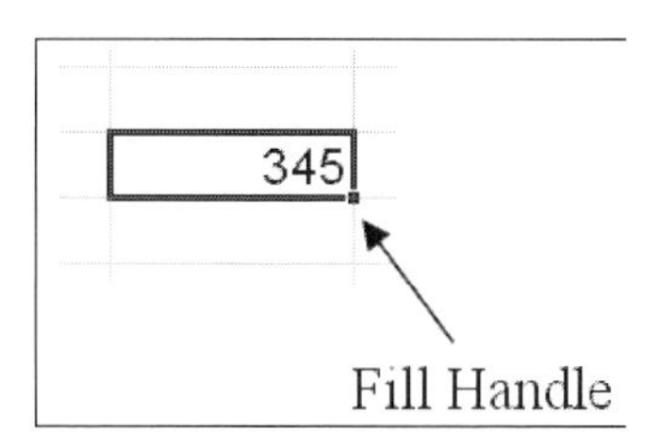

Figure 2.4. Example of a Fill Handle.

Figure 2.5. The Fill Series drop-down menu.

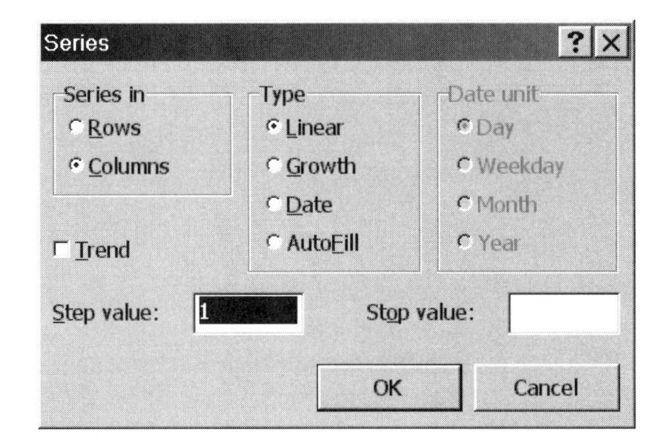

Figure 2.6. The Series dialog box.

PRACTICE!

Practice using Fill Handles and creating Fill Series by performing the steps that follow. Note that sometimes you will be using the left mouse button and sometimes you will be using the right mouse button.

First, practice copying the contents of a cell:

Step 1: Open a new worksheet.

Step 2: Type the value **1.5** into cell A1. Grab the Fill Handle for cell A1 with the left mouse button and drag it over cells B1:G1.

Step 3: When you release the mouse button, the value 1.5 will have been copied to cells B1:G1.

Now try creating a linear series:

Step 4: Type the values **1.5** and **1.7** into cells A3 and B3, respectively.

Step 5: Select the region A3:B3 and grab the Fill Handle with the right mouse button.

Step 6: Drag the fill handle over the cells C3:G3. When you release the mouse button, a linear series will have been created.

Note that Excel has created a linear Fill Series by using the difference between the first two cells as the increment. You can control the increment value in the following manner:

Step 7: Type the value **1.5** into cell A5.

Step 8: Grab the Fill Handle for cell A5 with the right mouse button and drag it over cells B5:G5.

Step 9: When you release the mouse button, the Fill Series drop-down menu will appear.

Step 10: Select **Series** from the drop-down menu. The Series dialog box will appear.

Step 11: Check **Rows** in the box labeled *Series in*.

Step 12: Check **Linear** in the box labeled *Type*.

Step 13: Type **0.4** in the box labeled *Step value*.

Step 14: Click **OK**.

The cells will be filled with a linear series in increments of 0.4. Your screen should now resemble Figure 2.7.

	A	B	C	D	E	F	G
1	1.5	1.5	1.5	1.5	1.5	1.5	1.5
2							
3	1.5	1.7	1.9	2.1	2.3	2.5	2.7
4							
5	1.5	1.9	2.3	2.7	3.1	3.5	3.9

Figure 2.7. Fill Series example.

2.5 FORMATTING CELLS

In the previous sections, you were shown how to format data types. Several other cosmetic formatting options are available for selected cells. These include the choice of fonts, colors, borders, shading, and alignment.

To access the cell formatting options, first select a region of cells and then choose **Format → Cells** from the Menu bar. An alternate method is to click the right mouse button and choose **Format Cells** from the Quick Edit menu. The Format Cells dialog box will appear.

From this dialog box, you may choose one of six tabs to format the font characteristics, select alignment within a cell, choose borders, colors, fill patterns, and enforce password protection of a worksheet. Choose the **Borders** tab. The Borders screen will appear, as shown in Figure 2.8.

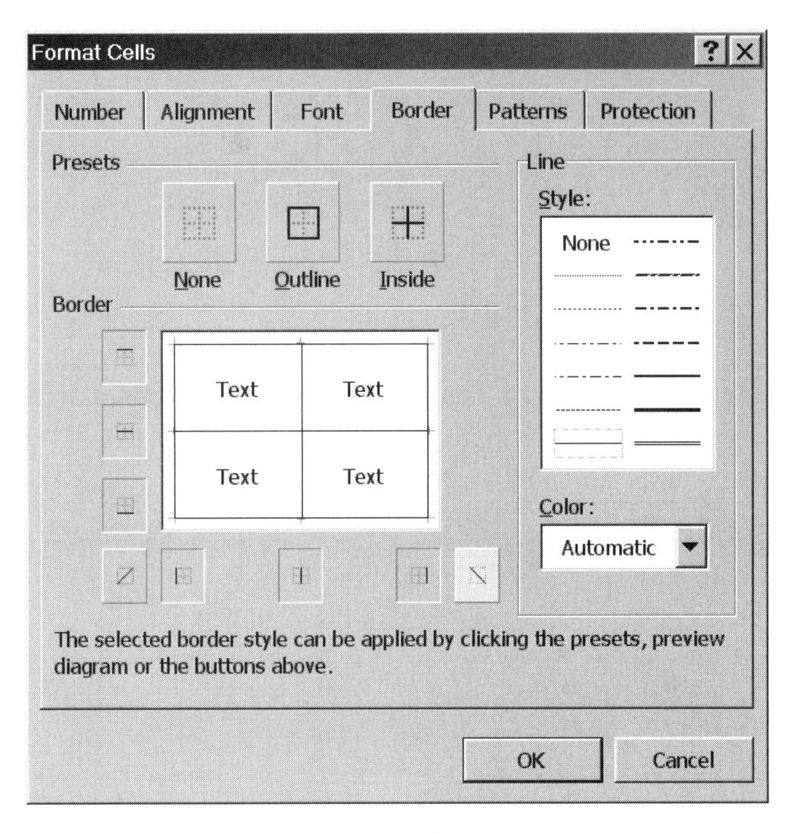

Figure 2.8. The Border screen of the Format Cells dialog box.

2.6 FORMATTING COLUMNS AND ROWS

The primary use of the row and column formatting option is to determine the row height and the column width. You can specify the height and width exactly or you can ask Excel to AutoFit the columns and rows for you. The *AutoFit* function adjusts the selected columns to the minimum width required to fit the data. If the data are changed, then the AutoFit may have to be reapplied.

A second use of the row and column formatting option is to hide or unhide a row or column. This may be useful for simplifying the view of a complex worksheet. To view the column formatting options, first select a region and then choose **Format → Column** from the Menu bar. A drop-down menu will appear, as shown in Figure 2.9.

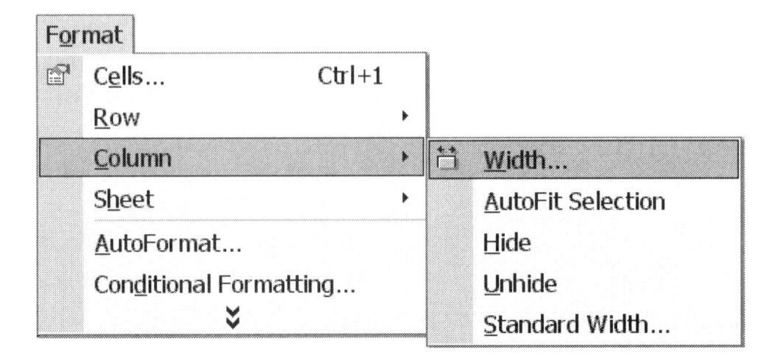

Figure 2.9. The Column Format drop-down menu.

From the Column Format menu, you can set the column width manually or choose the AutoFit function and let Excel determine the column width. You can set the default or standard width of a column by choosing **Standard Width** from the Column Format menu.

Similar formatting is available for rows by choosing **Format → Row** from the Menu bar.

2.7 TABLE AUTOFORMATS

Excel provides a few preformatted table types for convenience. To access the AutoFormat function,

Step 1: Select a region.

Step 2: Choose **Format → AutoFormat** from the Menu bar. The AutoFormat dialog box will appear as shown in Figure 2.10. Browse through the table formats in the list.

Step 3: When you have decided on a format, click **OK**.

2.8 CONDITIONAL FORMATTING

Formatting options may be applied conditionally. The term *conditional formatting* means you may choose logical criteria to format cells.

For example, consider the list of temperature data in Figure 2.11. Suppose that we want to emphasize the cells that show a station with temperature less than 20.0 degrees or greater than 51.1 degrees centigrade. We can use conditional formatting to highlight cells that meet those criteria.

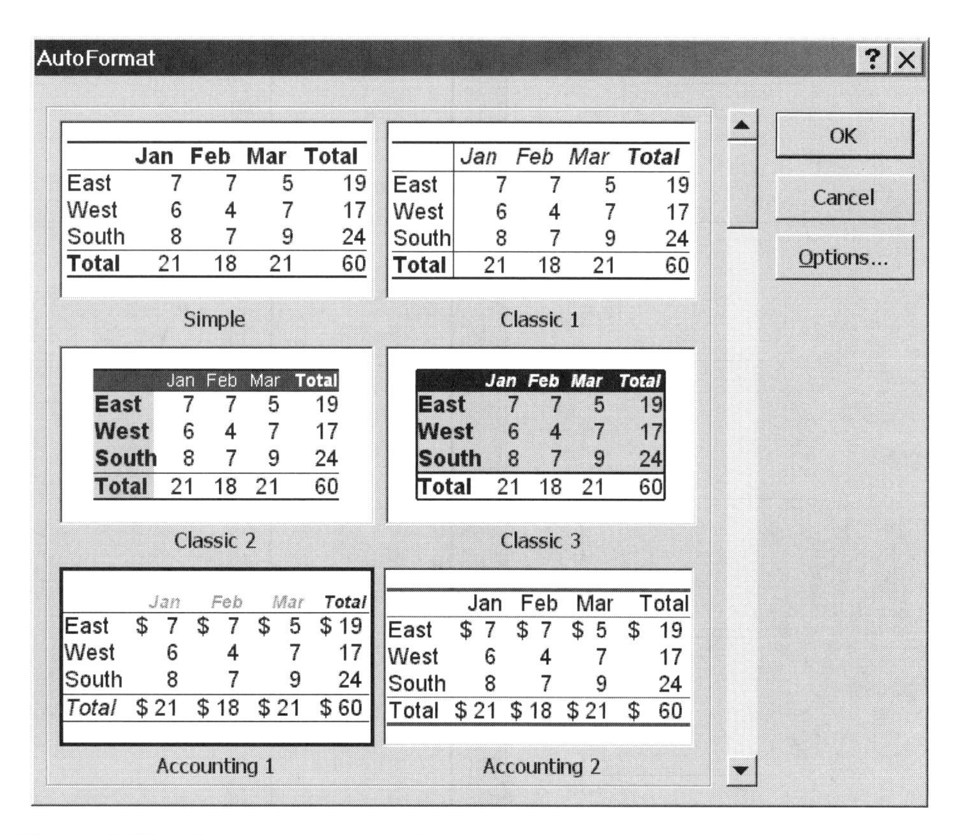

Figure 2.10. The AutoFormat dialog box.

	A	B
1	Station	Temp (C)
2	Newark	12.5
3	Santa Fe	23.4
4	Des Moines	53.2
5	Miami	72.6
6	Seattle	23.5
7	Phoenix	34.6

Figure 2.11. Temperature data from several monitoring stations.

Create the worksheet shown in Figure 2.11. To initiate conditional formatting, select the region B2:B7, then choose **Format → Conditional Formatting** from the Menu bar. The Conditional Formatting dialog box will appear.

To gain an understanding of how conditional formatting works, perform the following steps:

Step 1: Use the pull-down menus on the Conditional Formatting dialog box to set the criteria. For *Condition 1*, choose **Cell Value Is, less than**, and then type in **20.0**.

Step 2: Choose the **Format** button. The Format Cells dialog box will appear.

Step 3: Select the **Patterns** tab. Choose a cell shading color, then press **OK**.

Step 4: In the Conditional Formatting dialog box, choose the **Add** button.

Step 5: For *Condition 2*, choose **Cell Value Is, greater than**, and then type in **51.1**.

Step 6: Choose the **Format** button. The Format Cells dialog box will appear.

Step 7: Select the **Patterns** tab. Choose a cell shading color, then press **OK**.

Step 8: In the Conditional Formatting dialog box, choose the **Add** button.

Step 9: Your Conditional Formatting dialog box should resemble Figure 2.12. When you are satisfied with the entries, press **OK**.

Your worksheet should now resemble Figure 2.13. The cells with temperatures meeting the criteria you specified should now be highlighted.

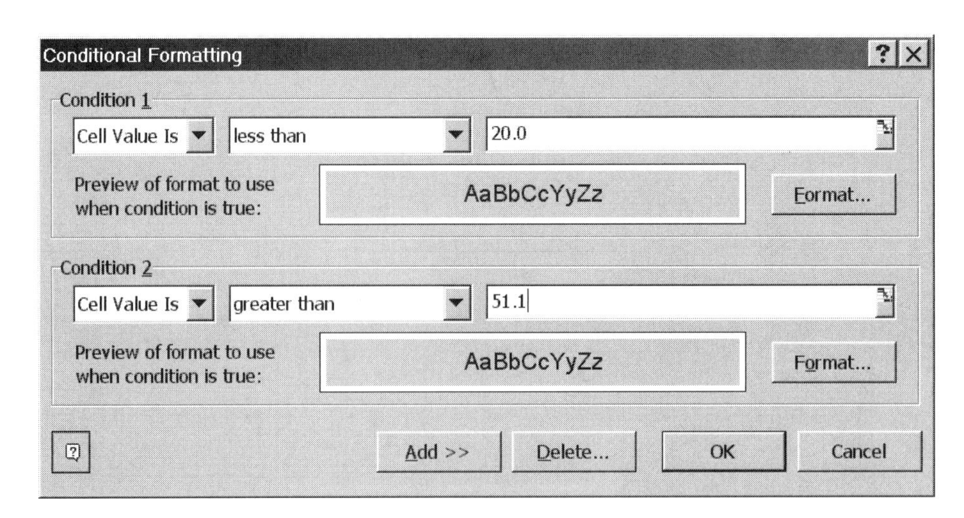

Figure 2.12. The Conditional Formatting dialog box.

	A	B
1	**Station**	**Temp (C)**
2	Newark	12.5
3	Santa Fe	23.4
4	Des Moines	53.2
5	Miami	72.6
6	Seattle	23.5
7	Phoenix	34.6

Figure 2.13. Results of applying conditional cell formats.

2.9 SORTING

The columns of a selected region may be sorted based on the values in the column. Columns can be sorted in ascending or descending order based on numeric, date, or alphabetic values. Multiple column sorts may be performed.

To sort the temperature data in Figure 2.13 alphabetically on the station name,

Step 1: Select column **A**.

Step 2: Choose the **Data → Sort** from the Menu bar. The Sort Warning dialog box will appear as shown in Figure 2.14. The warning tells you that there are nonempty cells next to our selection. If you sort only on column *A*, then the cities will no longer be associated with the correct temperatures. Since you want the temperatures to be sorted too, check the box labeled *Expand the selection* and then click **Sort**.

Step 3: The Sort dialog box will appear, as shown in Figure 2.15. Make sure that the column name **Station** has been selected in the box labeled *Sort by*.

Step 4: Check the box labeled *Ascending* and then click **OK**.

Your worksheet should now resemble Figure 2.16.

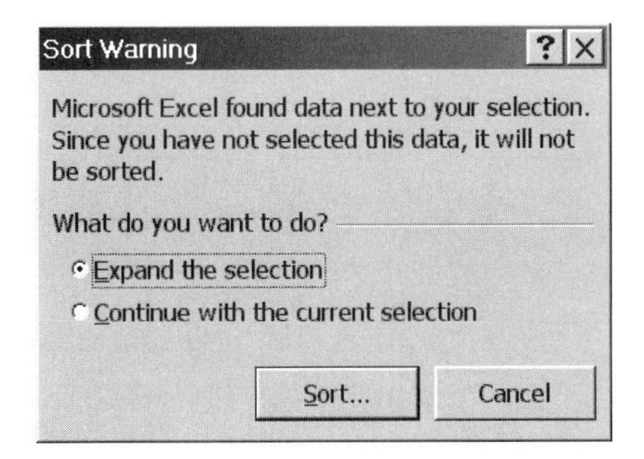

Figure 2.14. The Sort Warning dialog box.

Figure 2.15. The Sort dialog box.

	A	B
1	Station	Temp (C)
2	Des Moines	53.2
3	Miami	72.6
4	Newark	12.5
5	Phoenix	34.6
6	Santa Fe	23.4
7	Seattle	23.5

Figure 2.16. Example of an alphabetical sort.

PRACTICE!

Practice creating a multilevel sort as follows:

Create the table shown in Figure 2.17. The table in Figure 2.17 contains data about the melting and boiling points of elements called the noble gases. A noble gas is a non-metallic element that exists as a single atom in the table of elements.

	A	B	C
1	Symbol	Property	Temp (C)
2	He	Boil	-268.9
3	Xe	Melt	-112.0
4	Ne	Melt	-248.6
5	Ar	Boil	-185.8
6	Rn	Boil	-61.8
7	Ar	Melt	-189.3
8	Kr	Boil	-152.9
9	Kr	Melt	-157.0
10	Rn	Melt	-71.0
11	Ne	Boil	-245.9

Figure 2.17. Properties of the noble gases.

Suppose that you want to sort the table into two categories so that all of the melting points appear first and the boiling points appear last. Furthermore, within each category, the temperature should be listed in descending order. You should procede as follows:

Step 1: Select the region containing the table, A1:C11.

Step 2: Choose **Data** → **Sort** from the Menu bar. The sort dialog box will appear.

Step 3: In the *Sort by* box, choose the **Property** column. Check *Ascending* order.

Step 4: In the *Then by* box, choose **Temp (C)**. Check *Descending* order.

Step 5: Your Sort dialog box should resemble Figure 2.18. When you are satisfied with the contents, click **OK**.

Step 6: Your worksheet should now resemble Figure 2.19.

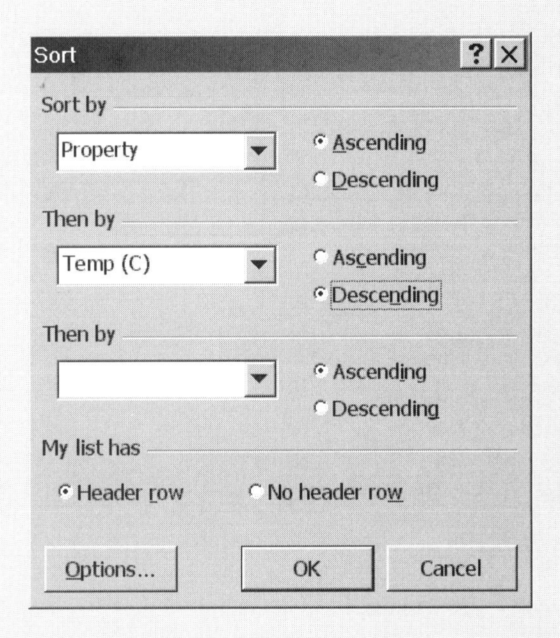

Figure 2.18. Creating a multilevel sort.

	A	B	C
1	Symbol	Property	Temp (C)
2	Rn	Boil	-61.8
3	Kr	Boil	-152.9
4	Ar	Boil	-185.8
5	Ne	Boil	-245.9
6	He	Boil	-268.9
7	Rn	Melt	-71.0
8	Xe	Melt	-112.0
9	Kr	Melt	-157.0
10	Ar	Melt	-189.3
11	Ne	Melt	-248.6

Figure 2.19. The results of a multilevel sort.

2.10 FORMATTING ENTIRE WORKSHEETS

Several formatting options apply to an entire worksheet. A worksheet can be hidden or renamed, or a different background can be selected. These functions can be accessed by choosing **Format → Sheet** from the Menu bar. The Format Sheet drop-down menu will appear, as shown in Figure 2.20.

A workbook can easily grow into a collection of dozens of worksheets. It is helpful to distinguish among worksheets by giving them meaningful names or by changing the background color or tab color of the worksheet. The default names (Sheet 1, Sheet 2, etc.) are not very helpful. Another strategy for organizing a large workbook is to hide worksheets that present underlying data, but keep visible worksheets that show results.

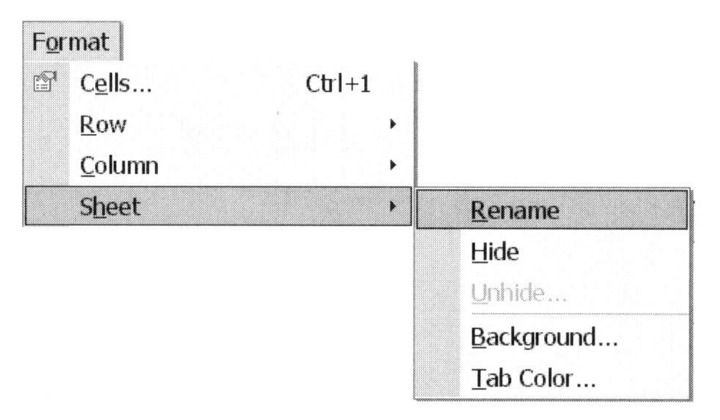

Figure 2.20. The Format Sheet drop-down menu.

APPLICATION—ENGINEERING ECONOMICS

Engineering economics involves the study of interest, cash flow patterns, and techniques for maximizing net value, depreciation, and inflation. This is an important area of study for all engineers, since engineers frequently serve as managers or executive officers of corporations.

The next example demonstrates how John can make a choice between investing $10,000 in a savings account that he knows will give him 6% growth or in the stock market in a fund that has historically shown 11% growth. John knows that there is no guarantee that the stock fund will continue to return a rate of 11%, but Excel allows him to forecast at 10-year intervals the growth differential between 6% and 11%, and he can use this information to decide if it would be worth it to take a chance that it will.

John's analysis shows that, if the stock market does return 11%, in 40 years, he will end up with over six times more money than if he puts the $10,000 into a savings account at 6%. John's results are displayed in Figure 2.21.

Accumulated capital		
Age	6% growth	11% growth
18	$10,000	$10,000
28	$17,908	$28,394
38	$32,071	$80,623
48	$57,435	$228,923
58	$102,857	$650,009

Figure 2.21. The results of John's analysis.

HOW'D HE DO THAT?

If John invested $10,000 in an account that paid 6% interest (6% APR, or *annual percentage rate*), one year later he would receive an interest payment of

$$0.06 \times \$10{,}000 = \$600$$

When the interest is added to the account, the value of the account after one year would be $10,000 + $600 = $10,600. To develop an equation that can be used to determine the value in the account at the end of any year, let's define some variables:

P = present value=the amount of John's initial deposit;
i = fractional interest rate (i.e., 0.06, not 6);
F = future value = the value at the end of any year;
N = number of years since the initial deposit.

After one year, the amount in John's account can be computed as follows:

$$F_1 = \$10{,}000 + (0.06 \times \$10{,}000) = P + iP = P[1 + i].$$

After a second year, the amount in the account will be as follows:

$$F_2 = F_1 + iF_1 = \{P[1 + i]\} + i\{P[1 + i]\} = P[1 + i][1 + i] = P[1 + i]^2.$$

After N years, the amount in the account will be as follows:

$$F = P[1 + i]^N.$$

The term $[1 + i]^N$ is called the *single payment compound amount factor*, and you can find tables of these factors for various interest rates in most economics texts, or you can easily create the table in Excel. A table created by using Excel is shown in Figure 2.22.

John calculated the future value of his money at 6% by using the compound amount factors in the 6% column. After 10 years, the $10,000 John invested would be worth 1.7908 times his initial investment:

$$F_{10} = \$10{,}000 \times 1.7908 = \$17{,}908.$$

After 20 years at 6%, the compound amount factor is 3.2071, so John's initial investment will have grown to

$$F_{20} = \$10{,}000 \times 3.2071 = \$32{,}071.$$

By using the compound amount factors table, it was easy for John to calculate the future value of his money for each of the two possible interest rates. John gathered his data into a worksheet as shown in Figure 2.23.

	A	B	C	D	E	F	G	H
1	Compound Amount Factors							
2								
3	Interest rate:		6%	7%	8%	9%	10%	11%
4	Fractional rate:		0.06	0.07	0.08	0.09	0.1	0.11
5								
6		Year						
7		0	1.0000	1.0000	1.0000	1.0000	1.0000	1.0000
8		10	1.7908	1.9672	2.1589	2.3674	2.5937	2.8394
9		20	3.2071	3.8697	4.6610	5.6044	6.7275	8.0623
10		30	5.7435	7.6123	10.0627	13.2677	17.4494	22.8923
11		40	10.2857	14.9745	21.7245	31.4094	45.2593	65.0009

Figure 2.22. Single-payment compound factors.

	A	B	C
1	Accumulated capital		
2	Age	6% growth	11% growth
3	18	10000	10000
4	28	17908	28394
5	38	32071	80623
6	48	57435	228923
7	58	102857	650009

Figure 2.23. John's first attempt.

HOW'D HE MAKE HIS TABLE LOOK SO GOOD?

John's first attempt at a table in Figure 2.23 contains the correct data, but the table could look better. Type the data in Figure 2.23 into a new worksheet and then complete the following steps to make your table look like the table in Figure 2.21:

Add Dollar Signs to the Values.

To add dollar signs to the values in columns B and C,

Step 1: Select the region B3:C7.

Step 2: Choose **Format Cells**. The Format Cells dialog box is displayed.

Step 3: Choose the **Number** tab, and select **Currency** format.

Step 4: Set the number of displayed decimal places to zero.

Step 5: See Figure 2.24. When you are satisfied with the selections, click **OK**.

Step 6: When the currency format has been applied, column **C** is no longer wide enough to display the last value. To have Excel adjust the column width to fit the contents, choose **Format → Column → Auto FitSelection**.

Center the Values.

To center the values in the Age column,

Step 1: Select the region A2:A7.

Step 2: Click on the **Center Alignment** icon from the Formatting toolbar. If the Formatting toolbar is not visible, choose **View → Toolbars** and check the box labeled *Formatting*. The *Age* heading and the age values should now be centered in column A.

Make the Column Headings Boldface.

To make column headings in row 2 boldface,

Step 1: Select the column headings in row 2 (region A2:C2)

Step 2: Choose the **Bold** icon on the Formatting toolbar.

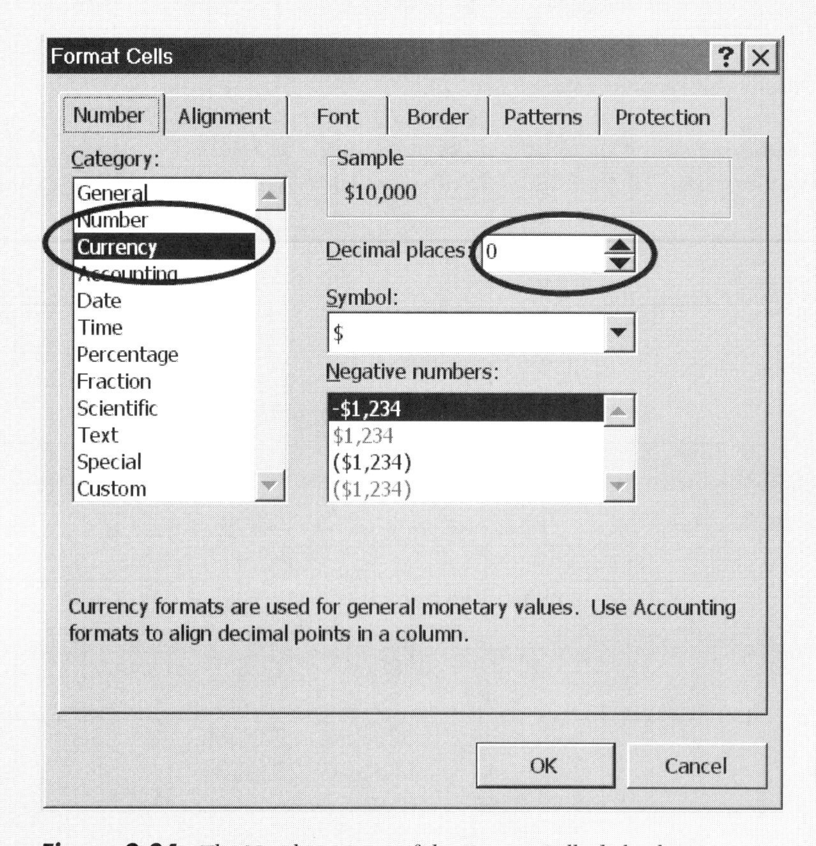

Figure 2.24. The Number screen of the Format Cells dialog box.

Step 3: Readjust the column widths by choosing **Format → Column → AutoFit Selection**.

Center the Title over the Entire Table

To center the title over the table, we will merge cells A1 through C1, and then center the new merged cell by following these steps:

Step 1: Select the region A1:C1.

Step 2: Choose the **Merge and Center** icon from the Formatting toolbar.

Add Lines Around Each Cell

To add lines around each cell, follow these steps:

Step 1: Select the region A1:C7.

Step 2: Choose **Format → Cells** and choose the **Border** tab.

Step 3: On the Border panel, choose a solid line from the box labeled *Line Style*.

Step 4: Click on the **Outline** and **Inside** buttons.

Step 5: See Figure 2.25. When you are satisfied with the selections, choose **OK**.

Change the Background and Text colors of the Title

Follow these steps to change the background and text colors of the titles,

Step 1: Select the title (the merged cells A1:C1).

Step 2: Choose **Format → Cells** and choose the **Patterns** tab.

Step 3: Choose a cell shading color. (Grey was selected in this example.)

Step 4: Select the **Font** tab.

Step 5: Set the *Font style* to **Bold**, and the *Color* to white.

Your completed table should now look like Figure 2.26.

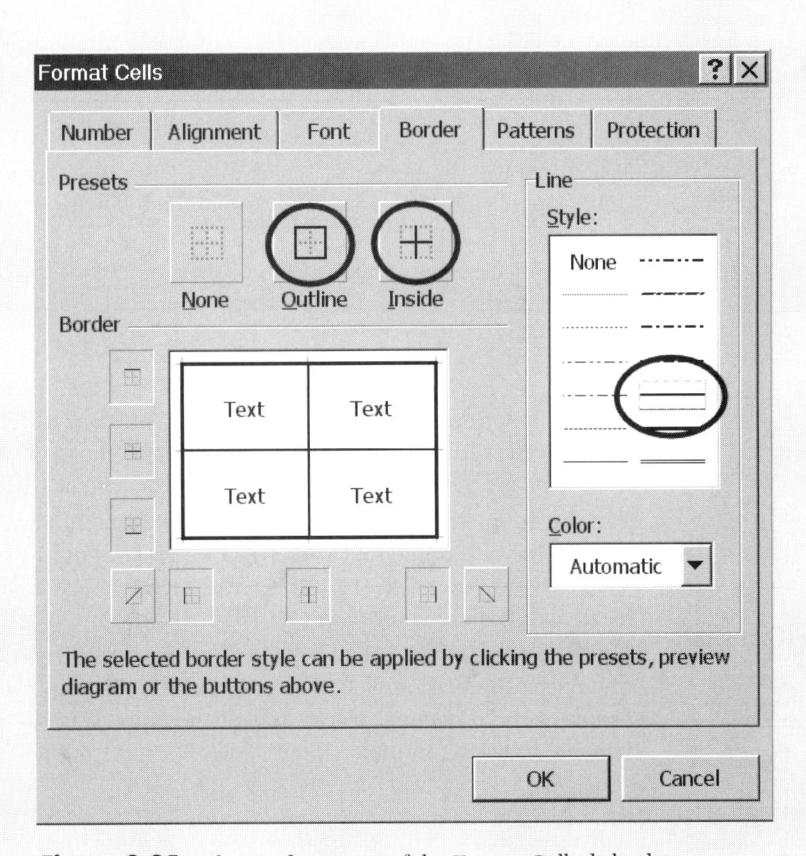

Figure 2.25. The Border screen of the Format Cells dialog box.

Accumulated capital		
Age	6% growth	11% growth
18	$10,000	$10,000
28	$17,908	$28,394
38	$32,071	$80,623
48	$57,435	$228,923
58	$102,857	$650,009

Figure 2.26. John's formatted table

KEY TERMS

AutoFit
Date format
General format
Time format
conditional formatting
Fill Handle
Scientific format
Currency format
Fill Series
Text format

SUMMARY

In this chapter, you were introduced to the data types used in Excel and the methods for entering data into a worksheet. The use of Fill Handles to quickly enter data was presented. You were also shown the commands for formatting cells, rows, columns, and worksheets. Finally, you were presented with methods for conditionally formatting cells and for sorting worksheets.

Problems

1. When you use the General format, Excel attempts to guess the type by your entry. Open a new workbook and type in the following items:

   1/2 _____
   1 1/2 _____
   $ 10.32 _____
   3E2 _____

   To what types does Excel automatically convert your text? To see the type, select the cell (with the data already entered) and choose **Format → Cells** from the Menu bar. The menu will be placed on the format type of the cell.
2. Use the on-line help features of Excel to determine how Excel deals with the Y2K problem. Assume that you are entering dates of birth in an Excel spreadsheet. If you enter 5/23/19, does Excel record the year to be 1919 or 2019? What about a date entered as 11/14/49?
3. Excel stores numbers with 15 digits of precision. Prove to yourself these limits of numerical precision. Select an empty cell and format the cell to the *Number* format with 20 decimal places. Select the equal sign on the formula bar and type the following formula:

   =SQRT(2).

   Since the square root of two is an irrational number, the fractional part of its decimal representation has an unending number of nonrepeating digits. At what number of digits does Excel's accuracy stop?

4. The table in Figure 2.27 shows the temperatures recorded at several monitoring stations for the months of January and February 2002. The data are in no particular order. Create a worksheet that looks like Figure 2.27. Sort the data in your worksheet by station and then by date in ascending order. The results should look like Figure 2.28.

	A	B	C
1	Station	Date	Temp (C)
2	A02	02/01/02	23.5
3	A07	01/15/02	14.6
4	A02	02/15/02	0.5
5	B05	02/03/02	20.0
6	B05	01/12/02	34.3
7	C12	01/05/02	20.2
8	A02	02/23/02	19.6
9	B05	02/01/02	22.3

Figure 2.27. Temperature data.

	A	B	C
1	Station	Date	Temp (C)
2	A02	02/01/02	23.5
3	A02	02/15/02	0.5
4	A02	02/23/02	19.6
5	A07	01/15/02	14.6
6	B05	01/12/02	34.3
7	B05	02/01/02	22.3
8	B05	02/03/02	20.0
9	C12	01/05/02	20.2

Figure 2.28. Sorted temperature data.

5. Use conditional formatting to highlight temperatures less than or equal to 20.0 degrees Celsius. The results should shade the same cells as shown in Figure 2.29.

	A	B	C
1	Station	Date	Temp (C)
2	A02	02/01/02	23.5
3	A02	02/15/02	0.5
4	A02	02/23/02	19.6
5	A07	01/15/02	14.6
6	B05	01/12/02	34.3
7	B05	02/01/02	22.3
8	B05	02/03/02	20.0
9	C12	01/05/02	20.2

Figure 2.29. Sorted data highlighting temperatures $\leq$ 20.0 C.

6. Create a worksheet containing the table of noble gases shown in Figure 2.17. Sort the table so that the elements are listed in alphabetical order. For each element, place the melting point first, followed by the boiling point. Your results should look like Figure 2.30.

	A	B	C
1	Symbol	Property	Temp (C)
2	Ar	Melt	-189.3
3	Ar	Boil	-185.8
4	He	Boil	-268.9
5	Kr	Melt	-157.0
6	Kr	Boil	-152.9
7	Ne	Melt	-248.6
8	Ne	Boil	-245.9
9	Rn	Melt	-71.0
10	Rn	Boil	-61.8
11	Xe	Melt	-112.0

Figure 2.30. Properties of the noble gases.

7. Use conditional formatting to highlight only the boiling points in the table of Figure 2.30. Your results should highlight the same items as in Figure 2.31.

	A	B	C
1	Symbol	Property	Temp (C)
2	Ar	Melt	-189.3
3	Ar	Boil	-185.8
4	He	Boil	-268.9
5	Kr	Melt	-157.0
6	Kr	Boil	-152.9
7	Ne	Melt	-248.6
8	Ne	Boil	-245.9
9	Rn	Melt	-71.0
10	Rn	Boil	-61.8
11	Xe	Melt	-112.0

Figure 2.31. Noble gases with boiling points highlighted.

8. Create and format a table that looks like Figure 2.32.

	A	B	C	D	E	F
1	Name	Quiz 1	Quiz 2	Midterm	Quiz 3	Final
2	Bob	23	12	43	21	54
3	Maria	32	10	40	26	55
4	Ralph	14	12	34	20	45
5	Deepak	24	13	38	22	58

Figure 2.32. Student grades.

3

Formulas and Functions

3.1 INTRODUCTION

The ability to manipulate formulas, arrays, and mathematical functions is the most important feature of Excel for engineers. A common scenario is for an engineer to test and refine potential solutions to a problem by using Excel. When the engineer is satisfied that the solution works for small data sets, the solution might be translated to a programming language such as C or FORTRAN. The resulting program could be then executed on a powerful workstation or supercomputer, using large data sets. This use of a worksheet is called building a *prototype*. An application package such as Excel is useful for building prototypes because it allows solutions to be quickly developed and easily modified.

In the next sections, formulas and functions will be used to solve two types of problems that should be familiar to engineering students: (1) finding the solutions to a quadratic equation and (2) matrix multiplication. These examples will be used to demonstrate many of the features of Excel that are related to engineering computation.

In addition, in this chapter, you will be introduced to Excel macros—a method for recording and executing a series of actions. The use of macros can be a time-saving feature as you learn to solve problems that require a series of computations.

SECTIION

OBJECTIVES

After reading this chapter, you should be able to:

- Create formulas in a worksheet.
- Use absolute and relative cell references.
- Locate and use Excel's predefined functions.
- Debug worksheet formulas that contain errors.
- Perform simple matrix operations with Excel.
- Record and run a macro.

3.2 CREATING AND USING FORMULAS

A formula in Excel consists of a mathematical expression. For the most part, the expression is defined by using common arithmetic operators. However, as we will learn later, there are several differences between mathematical symbols and Excel's operators.

The cell containing a formula can display either the formula definition or the results of applying the formula. The default is to display the results in the cell. This is usually preferable, since the formula definition for the currently active cell is displayed in the Formula bar.

Figure 3.1 shows cell D1 to be the currently active cell. The formula bar shows the formula definition of D1 as

= A1 * B1 / C1.

A formula definition begins with an equals sign (=). The result of applying this expression is displayed in cell D1 as *1.746296*. In this chapter, we will show you how to build and debug formula definitions.

3.3 FORMULA SYNTAX

An Excel formula uses a strict syntax, which means that you must learn the rules for entering a formula and adhere strictly to those rules. A formula can consist of operators, predefined function names, cell references, and cell names. The use of each of these syntactic groups will be covered in this chapter.

A formula always begins with an equals sign (=). This symbol is an indicator for Excel to evaluate the expression that follows, instead of simply placing the contents of the expression in the cell. Try removing the equals sign and see what happens.

You can enter a formula into the Formula window on the Formula bar. Or you can enter the formula directly into the destination cell.

3.3.1 Arithmetic Operators

Table 3.1 displays Excel's arithmetic operators. The operators are listed in order of precedence. For example, exponentiation will be calculated before addition. Operators in the same level, such as addition and subtraction, will be calculated from left to right. If a different precedence is desired, then parentheses must be used. There are other operators for the manipulation of text and for Boolean comparison that will not be covered here. Please use the on-line help for further information about these operators.

D1 ▾ fx =A1 * B1 / C1

	A	B	C	D
1	2.3	4.1	5.4	1.746296
2				
3				

Figure 3.1. An example of a formula.

TABLE 3.1 Arithmetic Operators

Precedence	Operator	Operation
1	%	Percentage
2	^	Exponentiation
3	*,/	Multiplication, Division
4	+, −	Addition, Subtraction

PRACTICE!

Practice creating arithmetic expressions by following these instructions:

Step 1: Select a cell.

Step 2: Type the following arithmetic expression into the cell:

= 6/2 + 3.

Try to figure out what the result will be before pressing the **Enter** key.

Step 3: Do the same for the following expressions:

= 6/(2 + 3);
= 2^2 − 1;
= 2^(2 − 1);
= 2 + 4/2/2 − 4.

3.3.2 Predefined Functions

A variety of predefined functions may be selected directly by choosing the **Insert Function** icon from the Formula bar. Choose the **Insert Function** icon. The Insert Function dialog box will appear, as shown in Figure 3.2.

From the Insert Function dialog box, you can select functions using several methods. One method is to type a topic in the box labeled *Search for a function*. Another method is to select a category and then a function from the drop-down menu. Once you have used several functions, a quick way to access them again is to select the **Most Recently Used** item in the category menu.

3.3.3 Cell References

Cell references can be entered into a formula in two ways. A cell location or range can be typed into the formula, or the cell range can be selected by using the mouse. Let's walk through an example that uses both methods of referencing cells:

Step 1: Type the values **7.5** and **6.2** into cells A3 and B3, respectively.

Step 2: Select cell C3.

Step 3: Choose the **Insert Function** icon from the Formatting toolbar. The Insert Function dialog box will appear.

Step 4: Choose the **Sum** function from the drop-down menu.

Step 5: The Function Arguments dialog box will appear, as shown in Figure 3.3.

Step 6: You have two choices at this point. You can type the input cell range into the box labeled *Number 1*, or you can select the small table icon that is circled in Figure 3.3. If you select the icon, the Function Arguments box will shrink, as shown in Figure 3.4. You can now select the input cell range by using the mouse. The selected region will have a dashed line around it. When you are satisfied with the selection, click on the small table icon (circled in Figure 3.4).

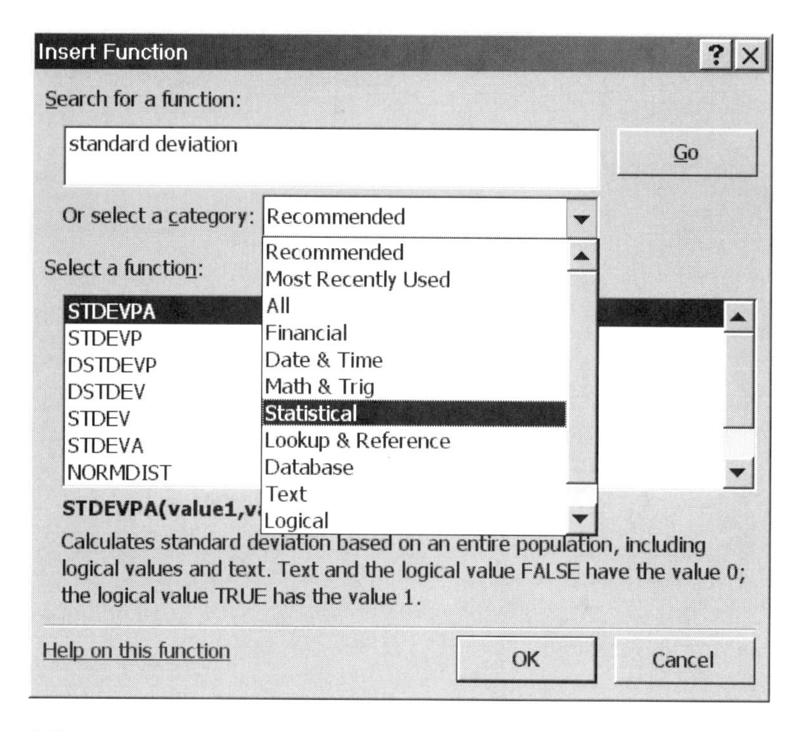

Figure 3.2. The Insert Function dialog box.

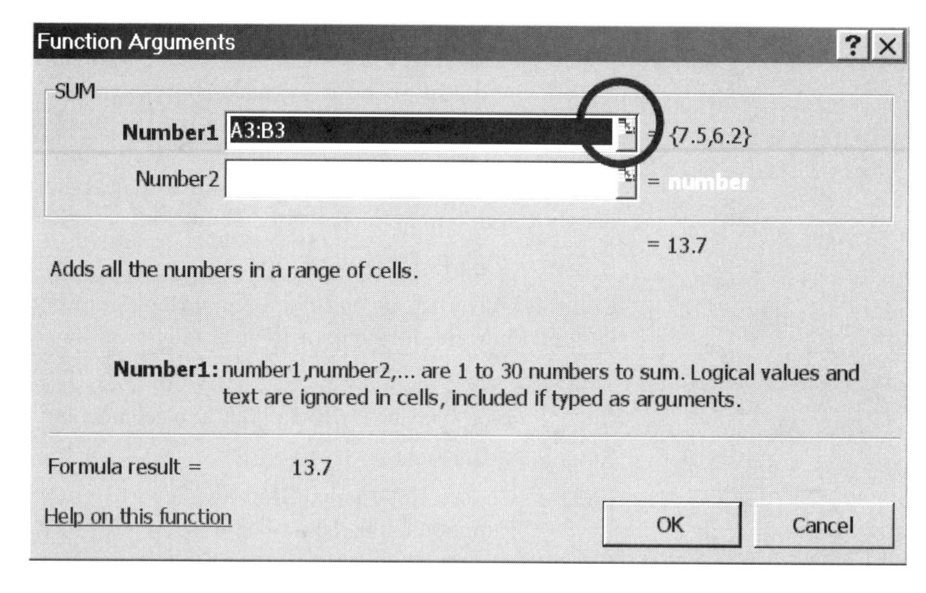

Figure 3.3. The Function Arguments dialog box.

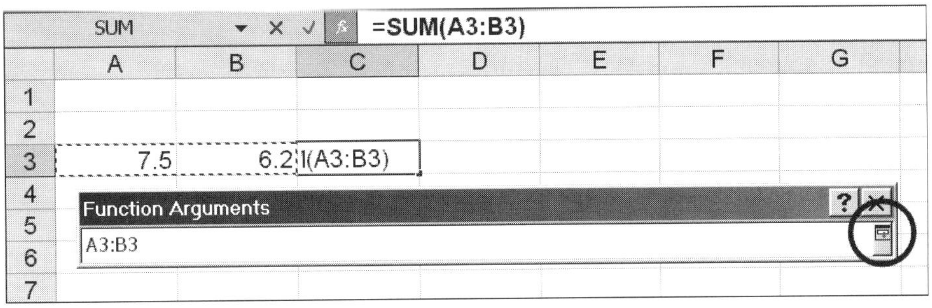

Figure 3.4. Selecting an input cell range with the mouse.

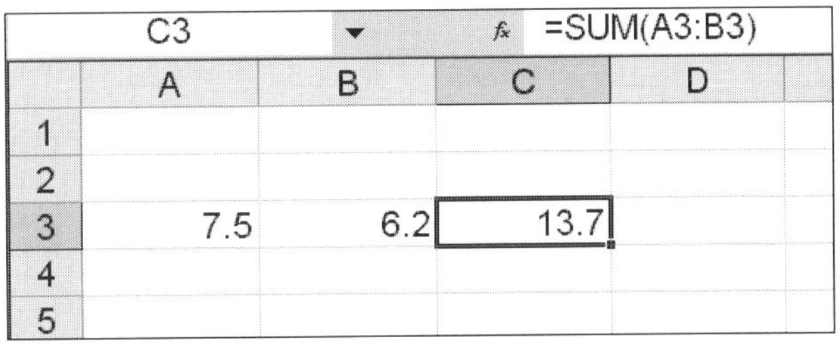

Figure 3.5. Sample formula results.

Step 7: The large version of the Function Arguments dialog box will now reappear. Note that the results of evaluating the formula are immediately displayed in the dialog box as you build the formula.

Step 8: Click **OK**. The results should resemble Figure 3.5.

Be careful not to insert a range of cells into an expression that doesn't make sense mathematically. For example, the formula

= SUM (A3:B3)

will sum cells A3 to B3. But the formula

= SQRT (A3:B3)

makes no sense, since the square root function accepts only one argument. You can't use SQRT to take the square root of a range of numbers. When an invalid range is entered, Excel responds by placing the following error message in the target cell:

#VALUE.

3.4 CELL AND RANGE NAMES

A group of cells can be given a name and the name can be added to a stored list. The name can then be used to reference the group of cells in a formula. Consider a range of cells that represents the following matrix:

$$A = \begin{bmatrix} 1 & -1 & 2 \\ 4 & 0 & -1 \\ -8 & 2 & 2 \end{bmatrix}.$$

	A	B	C
1	1	-1	2
2	4	0	-1
3	-8	2	-2

Figure 3.6. A 3 × 3 matrix with the diagonal elements selected.

This matrix can be represented in a worksheet, as shown in Figure 3.6. Assume that you want to perform operations on the *diagonal* of the matrix in a formula. For example, the following formula will sum the diagonal elements:

= SUM (A1, B2, C3).

These steps demonstrate how to name the diagonal elements of the matrix in Figure 3.6:

Step 1: Select the diagonal cells. Click on cell A1. Hold down the **Ctrl** key and click on cell B2. Hold down the **Ctrl** key and click on cell C3.

Step 2: Choose **Insert** → **Name** → **Define** from the Menu bar. The Define Name dialog box will appear, as shown in Figure 3.7.

Step 3: Type in a name for the group of cells (e.g., **Diagonal**). Your screen should resemble Figure 3.7.

Step 4: Click **OK** to finish the operation.

The name *Diagonal* is now associated with the cell range (A1, B2, C3) in this workbook. The name can be used anywhere the range is referenced within this workbook.

To use a previously defined name in a formula,

Step 1: Place the cursor at the insertion point.

Step 2: Choose **Insert** → **Name** → **Paste**.

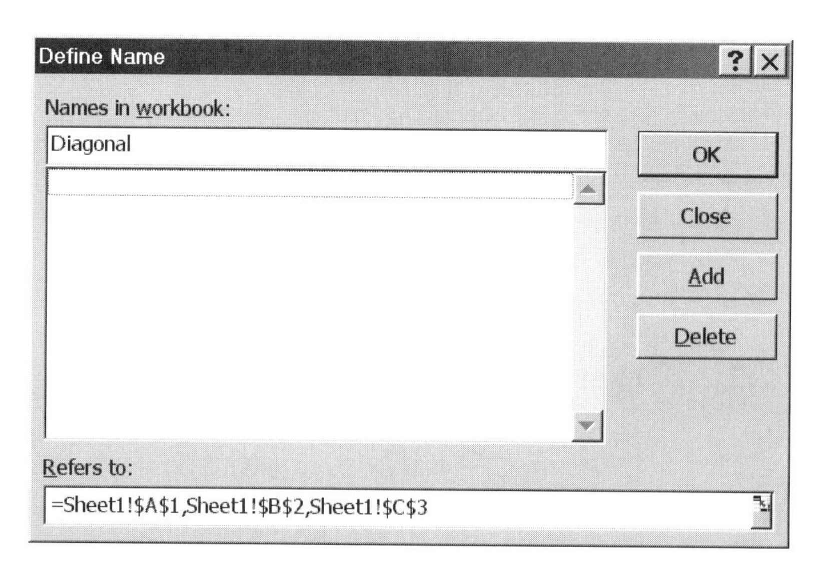

Figure 3.7. The Define Name dialog box.

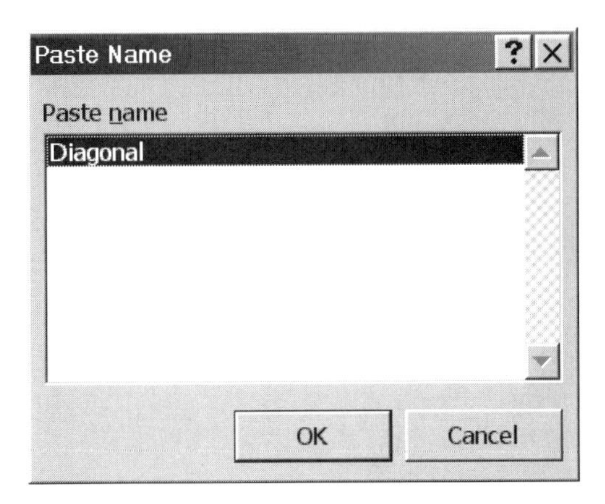

Figure 3.8. The Paste Name dialog box.

Step 3: The Paste Name dialog box will appear, as shown in Figure 3.8. Select a name (e.g., Diagonal) from the list.

Step 4: Click **OK**.

For example, a formula that sums the diagonal elements of the matrix can now be stated as

= SUM (Diagonal),

instead of

= SUM(A1, B2, C3).

This is more readable and less prone to typographical errors than the original version of the formula. The importance of using names will become clear as you begin to build more complex formulas and attempt to debug errors in them.

3.5 ABSOLUTE AND RELATIVE REFERENCES

Formulas may be copied from one location to another in a worksheet. Usually, you will want the cells that are referenced in the formula to follow the formula.

For example, you may want to copy a formula that adds a row of numbers. Look at Figure 3.9. The formula in cell E1 adds the elements in row 1 from A to C. If you were to copy the formula in E1 to E2, what effect would you want? Should the new formula compute the sum of row 1 or row 2?

If relative referencing is used, the copied formula in F2 will sum row 2 from columns A to C. Try the following test:

Step 1: Create the worksheet shown in Figure 3.9.

Step 2: Select cell E1.

E1 | =SUM(A1:C1)

	A	B	C	D	E
1	2	4	6	Sum=	12
2	3	5	7	Sum=	

Figure 3.9. Example of a formula with relative referencing.

Step 3: Choose **Edit → Copy** from the Menu bar. (You can also right click and choose **Copy** from the Quick Edit menu.)

Step 4: Select cell F1.

Step 5: Choose **Edit → Paste** from the Menu bar. (You can also right click and choose **Paste** from the Quick Edit menu.)

Note that the formula in cell F1 adds row 2, not row 1. The cell references have *followed* the formula. This is called *relative referencing*.

There are times when you may wish to copy a formula and *not* have the references follow the formula. This is called *absolute referencing*. A cell or range is denoted as an absolute reference by the placement of a dollar sign ($) in front of the row or column to be locked.

Look at Figure 3.10. This worksheet is the same as the worksheet in Figure 3.9, except that the formula uses absolute references. Copy cell E1 to cell F1 and note the results. The formula in cell F1 sums row 1, not row 2! The references have **not** followed the formula.

E1 | fx =SUM($A$1:$C$1)

	A	B	C	D	E
1	2	4	6	Sum=	12
2	3	5	7	Sum=	

Figure 3.10. Example formula with absolute referencing.

PRACTICE!

There are times when you may want to make relative references to some cells and absolute references to other cells. For example, a formula may use a constant and several variables. You can use absolute referencing for the constant and relative referencing for the variables.

In the next example, we want to compute the area of a circle for several radii. The formula uses the constant pi. When we copy the formula, we want the reference to pi to remain absolute, but the reference to the radius to be relative. Complete the following steps:

Step 1: Create a new worksheet like the worksheet in Figure 3.11.

Step 2: Create a formula for cell B4, as shown in the Formula bar. Note that cell C1 uses an absolute reference and cell A4 uses a relative reference.

Step 3: Copy the formula in cell B4 by dragging the Fill Handle through cells B5:B8.

Step 4: The areas of the circles with radii A5:A8 should be correctly computed and displayed in cells B5:B8.

B4 | fx =$C$1 * A4^2

	A	B	C	D
1		pi=	3.14159	
2				
3	Radius	Circumference		
4	1.00	3.14		
5	4.23			
6	3.20			
7	18.60			
8	123.43			

Figure 3.11. Example of combined relative and absolute referencing.

PRACTICE!

An example of a slightly more complex use of formulas is the solution of quadratic equations. You may recall from high school algebra that if a quadratic equation is expressed in the form

$$ax^2 + bx + c = 0,$$

then the solutions for x are as follows:

$$x = -b \pm \frac{\sqrt{b^2 - 4ac}}{2a}, \quad (2a \neq 0)$$

Since Excel does not directly recognize imaginary numbers, we must make the further restriction that

$$b^2 - 4ac \geq 0.$$

Create a new worksheet that resembles Figure 3.12. Store coefficients a, b, and c in columns A, B, and C respectively. Place the formulas for the two solutions in D2 and E2, respectively. The Excel formula for the first solution is

= (−B2 + SQRT (B2^2 − 4 * A2 * C2)) / (2 * A2)

The formula for the second solution is

= (−B2 − SQRT (B2^2 − 4 * A2 * C2)) / (2 * A2)

Use the Fill Handle to copy the formulas for Root 1 and Root through rows 3 to 6. Your results should resemble Figure 3.13.
Notice how the row numbers in the formulas change, as you are using relative references. Also, notice that an error is displayed for the solutions on row 5, since one of our assumptions is violated ($2a = 0$).

D2 | =(-B2+SQRT(B2^2-4*A2*C2))/(2*A2)

	A	B	C	D	E
1	Coefficient A	Coefficient B	Coefficient C	Root 1	Root 2
2	1	2	1	-1	-1
3	1	16	1		
4	-4	-8	24		
5	0	18	24		
6	-4	-8	4		

Figure 3.12. Solutions for quadratic equations (partial).

H13 |

	A	B	C	D	E
1	Coefficient A	Coefficient B	Coefficient C	Root 1	Root 2
2	1	2	1	-1.00	-1.00
3	1	16	1	-0.06	-15.94
4	-4	-8	24	-3.65	1.65
5	0	18	24	#DIV/0!	#DIV/0!
6	-4	-8	4	-2.41	0.41

Figure 3.13. Solutions for quadratic equations (complete).

3.6 ERROR MESSAGES

The formulas that we have presented so far are relatively simple. If you make an error when typing one of the example formulas, the location of the error is relatively easy to spot. As you begin to develop more complex formulas, locating and debugging errors becomes a more difficult problem.

When a syntactic error occurs in a formula, Excel will attempt to immediately catch the error and then display an error box that explains the error. However, formulas can be syntactically correct, but still produce errors when the formula is executed. If an expression cannot be evaluated, then Excel will denote the error by placing one of eight error messages in the target cell. These error messages are listed in Table 3.2.

TABLE 3.2 Excel Error Messages

######	The value is too wide to fit in the cell, or an attempt was made to display a negative date or time.
#VALUE	The wrong type of argument was used in a formula. This will occur, for example, if text were entered when an array argument was expected.
#DIV/0	An attempt was made to divide by zero in a formula. See the quadratic equation example in Figure 3.13 for an illustration.
#NAME	A name used in a formula is not recognized. Usually the function or defined name was misspelled. Note that named ranges or functions may not contain spaces.
#REF	A referenced cell is not valid. This usually occurs when a cell is referenced in a formula and that cell is then deleted. It also occurs if an attempt is made to paste a cell over a referenced cell.
#NUM	The expression produces a numeric value that is out of range or invalid. Examples are extremely small, large, or imaginary numbers. Try this formula: = SQRT(−1).
#NULL	An attempt was made to reference the intersection of two areas that don't intersect. This usually occurs when a space is inadvertently placed between two arguments, instead of a comma or colon, as in = SUM(C2 D3).

3.7 DEBUGGING ERRORS BY USING CELL SELECTION

Debugging errors in a worksheet is made much easier by the use of the special cell selection option. To view the special cell selection menu,

Step 1: Select a range of cells to view.

Step 2: Choose **Edit → Go To** from the Menu bar. The Go To dialog box will appear.

Step 3: Choose the **Special** button. The Go To Special dialog box will appear, as shown in Figure 3.14.

From the Go To Special dialog box, a variety of options may be used to assist in the debugging process. We will discuss several of the options that are most relevant to debugging mathematical formulas.

3.7.1 Formulas

A common worksheet error is the accidental replacement of a formula with a constant. If you check the box labeled *Formulas*, all cells that contain formulas will be highlighted. You can further refine this option by selecting formulas that result in numerical, text, logical, or error values.

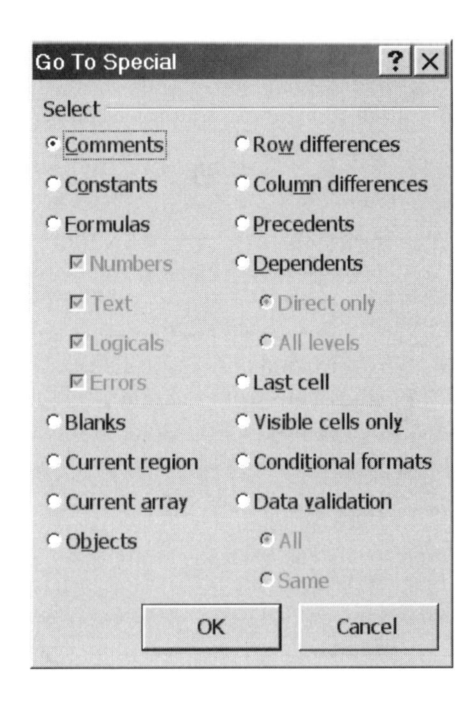

Figure 3.14. The Go To Special dialog box.

D3 =A3*B3

	A	B	C	D
1				
2				
3	3	4		12

Figure 3.15. Example of precedents and dependents.

3.7.2 Precedents

The *precedents* option displays cells that precede the selected cell(s). The displayed cells are all necessary for the computation of the selected cell(s). For example, in Figure 3.15, cell D3 contains the formula

= A3 * B3.

Cells A3 and B3 precede D4 and are said to be precedents of cell D4.

This option may be refined by checking or choosing the boxes labeled either *Direct Only* or *All Levels*. The Direct Only option will display the immediate precedents. The All Levels option will recursively display precedents (precedents of precedents).

3.7.3 Dependents

The *dependents* option displays cells that depend on the selected cell(s). For example, in Figure 3.15, if you select cell B3 and check the box labeled *Dependents* option, cell D4 will be displayed. This is because the formula in cell D4 depends on cell B3 for its computation. Cell D4 is said to be a dependent of cells A3 and B3.

This option may also be refined by checking or choosing the boxes labeled either *Direct Only* or *All Levels*. The Direct Only option will display the immediate dependents. The All Levels option will recursively display dependents (dependents of dependents).

3.7.4 Column (or Row) Differences

"A reference," according to the on-line Help, "identifies a cell or a range of cells on a worksheet and tells Excel where to look for the values or data you want to use in a formula. With references, you can use data contained in different parts of a worksheet in one formula or use the value from one cell in several formulas. You can also refer to cells on other sheets in the same workbook, and to other workbooks. References to cells in other workbooks are called links." When the cells in a column or row have different commands from the other cells, such as formulas, constants, etc., then they are said to have a different *reference pattern*. The column (or row) differences option highlights the cells in a column or row that have the different pattern from the other cells. For example, if all the selected cells in a column contain a formula, except one, which contains a constant, then the odd cell will be displayed.

The following example demonstrates the use of the cell-selection technique to locate and correct an error. Figure 3.16 illustrates a worksheet that computes the standard deviation of the student grades in column A. The mean grade is placed in cell E2. The difference between each grade and the mean grade is computed in column B by using the formula

=A1-$E$2.

This difference is squared in column C by using the formula

= B1^2.

The problem is that the answer is incorrect! The standard deviation should be 12.66, not 11.98. The error was found by using the following steps:

Step 1: Select column B by clicking on the column label.

Step 2: Choose **Edit → Go To → Special**.

Step 3: Check the box labeled *Column Differences*, then click **OK**.

The result is that cell B7 is highlighted. By looking in the formula bar, you can see that B7 incorrectly contains a constant value instead of a formula. Someone must have accidentally typed 8.4 into cell B7, overwriting the formula.

B7 fx 8.4

	A	B	C	D	E	F	G
1	Grades	Grade-Mean	Grade-Mean Squared		Mean	Sum of Differences Squared	Standard Deviation
2	32	0.7143	0.5102		31.2857	1003.9069	11.97561
3	14	-17.2857	298.7959				
4	52	20.7143	429.0816				
5	26	-5.2857	27.9388				
6	18	-13.2857	176.5102				
7	45	8.4000	70.5600				
8	32	0.7143	0.5102				

Figure 3.16. Using cell selection to debug a formula.

The debugging tools prove their worth as the worksheet gets larger and more complex. Consider trying to find the same error without the debugging tool if column A contained 3,000 grades instead of 7 grades!

3.8 DEBUGGING ERRORS USING TRACING

Excel also provides a visual method for tracing precedents, dependents, and cells with errors. The visual method is called *Formula Auditing*. The Formula Auditing tool may be accessed by choosing **Tools → Formula Auditing** from the Menu bar. The easiest way to manipulate the visual tool is to select **Show Auditing Toolbar** from the drop-down menu.

The Auditing toolbar contains buttons for tracing precedents, tracing dependents, and tracing cells with errors. The tool draws blue arrows that show the direction of precedence. Figure 3.17 demonstrates the effect of repeatedly choosing **Trace Precedents** from the Auditing toolbar when cell G2 is selected. The flow of the calculation is easily viewed when using this method. Note that cell B7 has no precedent, indicating a possible error condition.

3.9 USING EXCEL'S BUILT-IN FUNCTIONS

Excel has a large number of built-in functions that are similar to functions in a programming language. A function takes a specified number of arguments as input and returns a value. Excel functions are organized into function groups. The groups include database functions, financial functions, text functions, and date/time functions. Two groups that are of the most interest to engineers are the **Math & Trig** functions and the **Statistical** functions. We'll demonstrate a few functions in these function groups.

3.9.1 Examples of Statistical Functions

The following steps will walk you through the use of two simple statistical functions that compute the mean and median of a list of numbers:

Step 1: Enter the following 7 midterm grades into the range A2:A8:

32 68 93 87 75 96 82.

Step 2: Select the range A2:A8 and format the cells to be of type **Number** with zero decimal places.

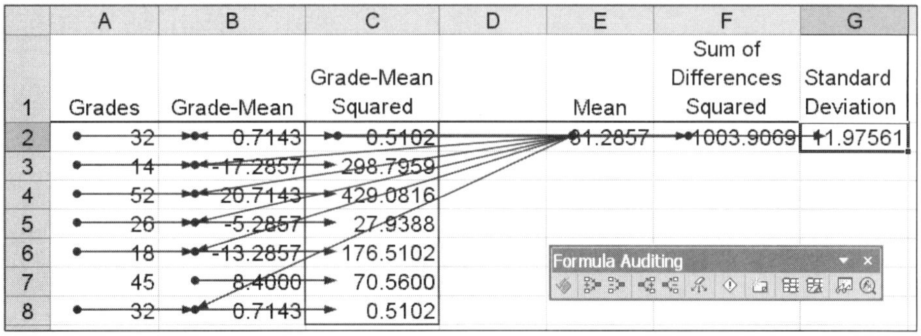

	A	B	C	D	E	F	G
1	Grades	Grade-Mean	Grade-Mean Squared		Mean	Sum of Differences Squared	Standard Deviation
2	32	0.7143	0.5102		31.2857	1003.9069	11.97561
3	14	-17.2857	298.7959				
4	52	20.7143	429.0816				
5	26	-5.2857	27.9388				
6	18	-13.2857	176.5102				
7	45	8.4000	70.5600				
8	32	0.7143	0.5102				

Formula Auditing

Figure 3.17. The Trace Precedents using the Formula Auditing tool.

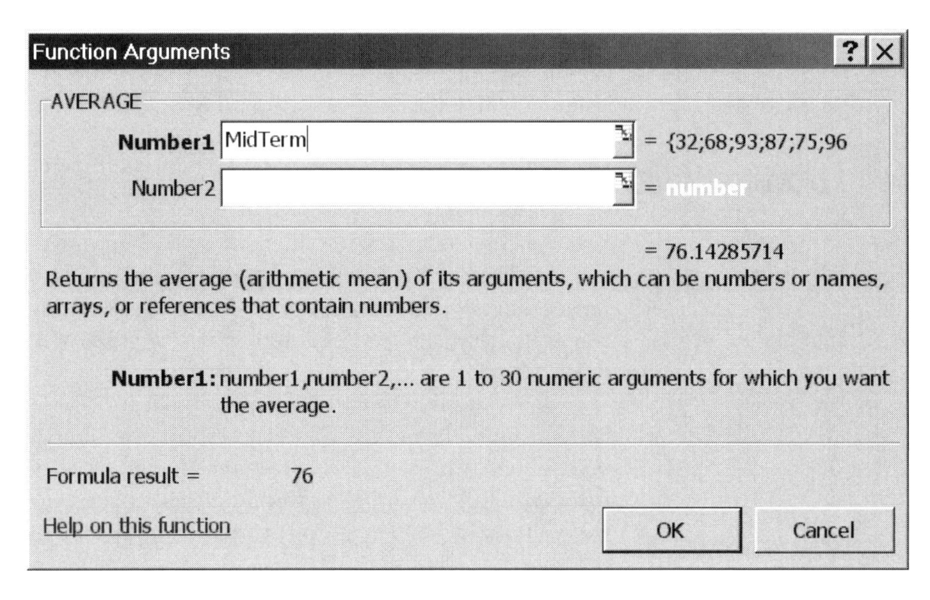

Figure 3.18. Computing the average midterm grade.

Step 3: Name the cell range **MidTerm** by selecting the region A2:A8 and then choosing **Insert → Name → Define** from the Menu bar. Type **MidTerm** and click **Add**. Then click **OK**.

Step 4: Select cell B2, which will hold the mean (average) of the midterm grades and then select the **Insert Function** icon from the Formula bar.

Step 5: The Insert Function dialog box will appear. Select **Statistical** from the category list and select **Average** from the function. Click **OK**.

Step 6: The Function Arguments dialog box will appear. Type **MidTerm** in the box labeled *Number 1*. The Function Arguments box should resemble Figure 3.18.

Step 7: Click **OK** to finish the operation.

PRACTICE!

Perform the preceding steps 1–7, except choose **Median** from the function-name list, instead of **Average**. Place the median grade in cell C2. Label each column appropriately. Your finished worksheet should resemble Figure 3.19.

C2 fx =MEDIAN(MidTerm)

	A	B	C	D	E
1	Midterm Grades	Mean Grade	Median Grade		
2	32	76.14	82.00		
3	68				
4	93				
5	87				
6	75				
7	96				
8	82				

Figure 3.19. The mean and median midterm grades.

PRACTICE!

Locate the Excel functions that compute the following mathematical functions:

1. trigonometric sine
2. inverse hyperbolic tangent
3. natural logarithm
4. base 10 logarithm
5. raise e to a power
6. convert radians to degrees
7. convert degrees or radians

3.9.2 Example of Matrix Operations

Matrices or arrays are frequently used in the formulation and solution of engineering problems. A *matrix* is a rectangular array of elements. Elements are referenced by row and column number. A spreadsheet is a natural application to represent and manipulate matrices. Excel has a number of built-in matrix operations that are included in the **Math & Trig** category of functions. These include the following:

MDETERM(array)— returns the matrix determinant for the named array;
MINVERSE(array)— returns the inverse of the named array;
MMULT(array1, array2)— performs matrix multiplication on the two named arrays.

There are also several predefined functions that compute sums or differences of products on matrices. In addition, many other functions take ranges as arguments and can be used to evaluate a matrix. For demonstration purposes, we'll use matrices A and B:

$$A = \begin{bmatrix} 3 & 1 \\ 4 & 3 \end{bmatrix}; \qquad B = \begin{bmatrix} 3 & -5 \\ 1 & 0 \end{bmatrix}.$$

You will now learn how to do matrix addition. Matrix addition is done by adding each of the corresponding cells of two matrices. The two matrices must be of the same order, which means that they both have the same number of rows and same number of columns. *Matrix order* is often denoted as the number of rows by the number of columns (rows × columns). The matrices in the previous example are of order 2 × 2. To add these two matrices, follow these simple steps:

Step 1: Enter matrices A and B with titles and borders as shown in Figure 3.20.

Step 2: Name matrices A and B by choosing **Insert → Name → Define** from the Menu bar.

Step 3: Select the region E2:F3.

	A	B	C	D	E	F
1	Matrix A		Matrix B		A + B	
2	3	1	3	-5		
3	4	3	1	0		

Figure 3.20. Matrices A and B.

E2 fx {=A+B}

	A	B	C	D	E	F
1	Matrix A		Matrix B		A + B	
2	3	1	3	-5	6	-4
3	4	3	1	0	5	3

Figure 3.21. Sum of matrices A and B.

Step 4: Type the formula

$$= \text{A} + \text{B}.$$

into the formula bar.

Now, you must indicate to Excel that you are asking for matrix addition. Simultaneously press the following keys: **Ctrl + Shift + Enter**. Curly braces will appear enclosing the formula. (*Note:* You cannot type the curly braces; the **Ctrl + Shift + Enter** key sequence must be used.) The range E2:F3 will now display the matrix sum of A and B. Your screen should resemble Figure 3.21.

As a second example, we will walk through the execution of the transpose operation. The *transpose* of a matrix is the matrix that is formed by interchanging the rows and columns of the original matrix. To find the transpose of B,

Step 1: Select the range C6:D7 to store the transpose of matrix B. The selected range must be of the same order as B (2×2).

Step 2: Type the following into the formula bar:

= TRANSPOSE(B).

Step 3: Press **Ctrl + Shift + Enter**. Curly braces will appear enclosing the formula.

Step 4: Place appropriate borders and labels on your worksheet. The results should resemble Figure 3.22.

C6 fx {=TRANSPOSE(B)}

	A	B	C	D	E	F
1	Matrix A		Matrix B		A + B	
2	3	1	3	-5	6	-4
3	4	3	1	0	5	3
4						
5			Transpose B			
6			3	1		
7			-5	0		

Figure 3.22. The transpose of matrix *B*.

PRACTICE!

Matrix multiplication is defined as follows: If $A = [a_{ij}]$ is an $m \times n$ matrix and $B = [b_{ij}]$ is an $n \times p$ matrix, then the *product* $AB = C = [c_{ij}]$ is an $m \times p$ matrix defined by

$$c_{ij} = \sum_{k=1}^{n} a_{ik}b_{kj}, \quad i = 1,2,\ldots,m, \quad j = 1,2,\ldots,p.$$

From this equation, the product of A and B is calculated to be

$$AB = \begin{bmatrix} 3 & 1 \\ 4 & 3 \end{bmatrix} \begin{bmatrix} 3 & -5 \\ 1 & 0 \end{bmatrix} =$$

$$\begin{bmatrix} (3\cdot 3) + (1\cdot 1) & (3\cdot -5) + (1\cdot 0) \\ (4\cdot 3) + (3\cdot 1) & (4\cdot -5) + (3\cdot 0) \end{bmatrix} = \begin{bmatrix} 10 & -15 \\ 15 & -20 \end{bmatrix}.$$

Excel has a built-in matrix multiplication function named **MMULT**. Use this function to verify the preceding results. You can practice using this function even if you have not yet studied matrix multiplication.

3.10 USING MACROS TO AUTOMATE COMPUTATIONS

A *macro* is a stored collection of commands. If you repeat the same set of commands repeatedly, then using a macro can be a convenient time-saving feature.

A macro is stored internally in a *Visual Basic module*. It is not within the scope of this text to teach you the Visual Basic language. However, Excel allows you to record and execute macros without knowing Visual Basic. In the next sections, you will be guided through recording and running a macro. Then you will be shown how to view the Visual Basic code that contains the macro commands. If you were to learn Visual Basic, then you could edit the code directly or write your own macros in the Visual Basic language.

3.10.1 Recording a Macro

Before recording a macro, it is wise to carefully plan the steps that you will be taking. When in recording mode, everything that you type is recorded—mistakes and all. In the next example, the major steps for computing several statistics of a set of data are listed. It is assumed that you are familiar with the use of Excel's built-in mathematical functions. If not, then review the previous sections in this chapter.

To record a sample macro, perform the following steps:

Step 1: Create a new worksheet and place

10 12 45 32 23 23 76 21 32 21

in cells A1:A10.

Step 2: Turn on macro recording by selecting **Tools → Macro → Record New Macro** from the Menu bar. The Record Macro dialog box will appear, as shown in Figure 3.23.

Step 3: Give your macro a name—the example uses **MyFirstMacro**. This dialog box also allows you to choose where to store the macro. Choose **This Workbook**. When you are done, press **OK**.

Note: Everything you now type will be recorded in the macro until you stop the recording process! To assist you in stopping the recording process, the small Stop Recording box will appear, as depicted in Figure 3.24.

Step 4: Select the region B2:E2 of the worksheet. Format the cells to be of type **Number** with two decimal points of accuracy.

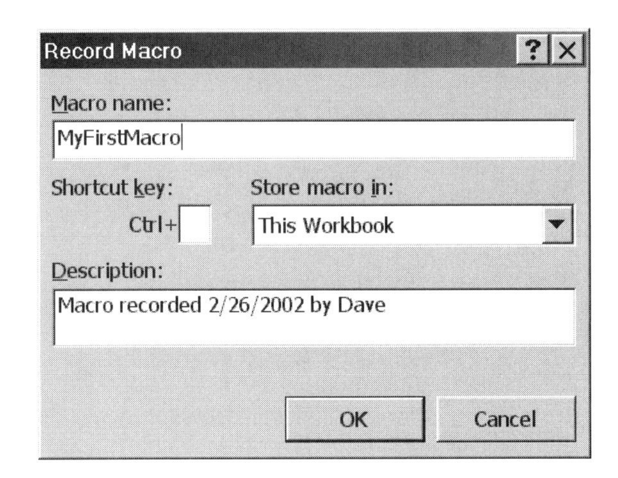

Figure 3.23. The Record Macro dialog box.

Step 5: Place the text **Mean** in cell B1, and place the formula AVERAGE(A1:A10) in cell B2.

Step 6: Place the text **Median** in cell C1, and place the formula MEDIAN(A1:A10) in cell C2.

Step 7: Place the text **Max** in cell D1, and place the formula MAX(A1:A10) in cell D2.

Step 8: Place the text **Min** in cell E1, and place the formula MIN(A1:A10) in cell E2.

Step 9: Center, and format your column labels.

Step 10: Press the **Stop** button in the Stop Recording dialog box. Congratulations, you have recorded a macro! Your worksheet should resemble Figure 3.25.

Step 11: Save your worksheet.

Figure 3.24. The Stop Recording dialog box.

	A	B	C	D	E
1	10	**Mean**	**Median**	**Max**	**Min**
2	12	29.50	23.00	76.00	10.00
3	45				
4	32				
5	23				
6	23				
7	76				
8	21				
9	32				
10	21				

Figure 3.25. Worksheet after recording a macro.

3.10.2 Executing a Macro

You can now retrieve and reuse your recorded macro at will. To see how powerful the use of macros can be, perform the following steps:

Step 1: Clear the contents of all of the cells on your current worksheet by choosing the top left button on the worksheet to select all cells. Then click the right mouse button and choose **Clear Contents** from the Quick Edit menu.

Step 2: Type 10 new numbers into cells A1:A10.

Step 3: Execute your macro by choosing **Tools** → **Macro** → **Macros** from the Menu bar. The Macro dialog box will appear as depicted in Figure 3.26.

Step 4: Select your macro's name from the list and then choose **Run**. Voila! Your worksheet will automatically perform all of the commands that you previously typed.

3.10.3 Editing a Macro

Several functions may be performed from the Macro dialog box. The macro may be deleted by choosing the **Delete** button. A shortcut key may still be added for a macro by choosing the **Options** button. The **Step Into** button allows you to debug a macro.

Select your macro name from the list and choose the **Edit** button on the Macro dialog box. The Microsoft *Visual Basic editor* (where you can create, modify, and manage your macros) will appear, as depicted in Figure 3.27. Inside the editor, you can view and modify the Visual Basic code that was created when you recorded the macro.

For example, the following two lines of code placed the text *Mean* in cell B1:

```
Range("B1").Select

ActiveCell.FormulaR1C1 = "Mean"
```

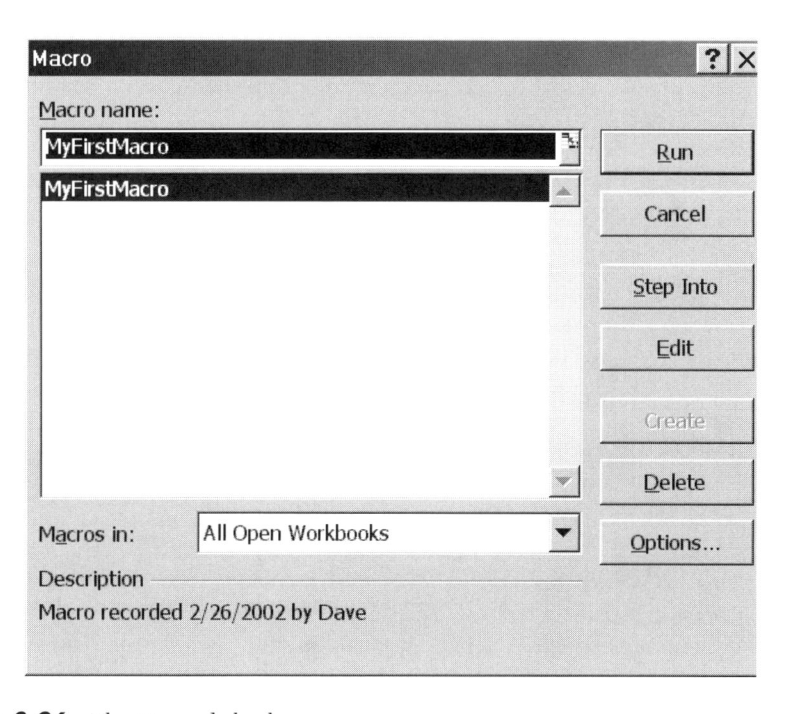

Figure 3.26. The Macro dialog box.

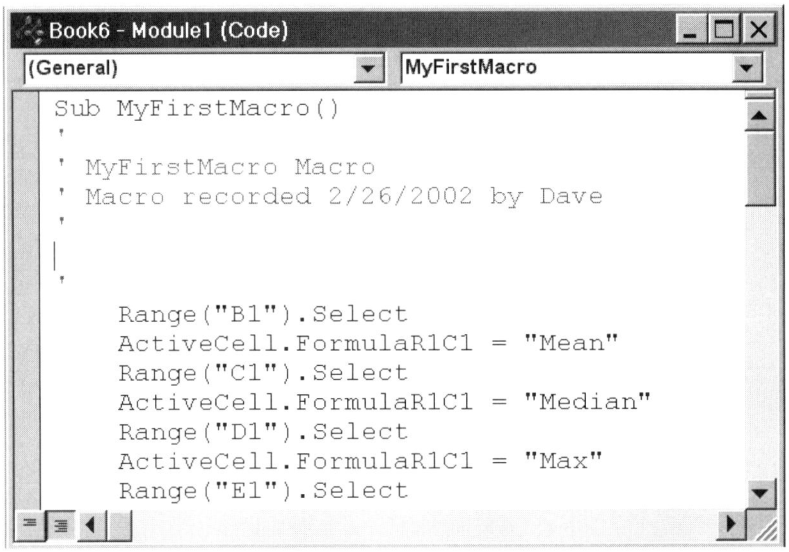

```
Book6 - Module1 (Code)
(General)                    MyFirstMacro
Sub MyFirstMacro()
'
' MyFirstMacro Macro
' Macro recorded 2/26/2002 by Dave
'

'
    Range("B1").Select
    ActiveCell.FormulaR1C1 = "Mean"
    Range("C1").Select
    ActiveCell.FormulaR1C1 = "Median"
    Range("D1").Select
    ActiveCell.FormulaR1C1 = "Max"
    Range("E1").Select
```

Figure 3.27. The Visual Basic editor.

PRACTICE!

Convince yourself that the code displayed in the Visual Basic editor is actually the same code contained in the macro. You can do this without knowing the Visual Basic language. Make several trivial changes, such as changing the line

ActiveCell.FormulaR1C1 = "Mean"

to

ActiveCell.FormulaR1C1 = "Average".

Execute the modified macro by choosing the **Run** button on the Visual Basic editor standard toolbar. Your worksheet should have changed to reflect your code change!

APPLICATION—CREATING TABLES OF COMPOUND AMOUNT FACTORS

Excel is a great tool for creating and displaying tables of information based upon repetition of formulas. The next example shows how to create a table that shows the power of compound interest—a compound amount factor table. By using this table, an investor can see how the value of an initial investment will increase over time at different interest rates. For example, a person who invests $1,000 at 6% interest for 20 years will have 1,000 × 3.2071 or $3,207. A person who invests $1,000 at 11% interest for 20 years will have 1,000 × 8.0623 or $8,062.

Because investments with higher interest payments normally carry a higher degree of risk, such a table helps one assess whether the potential increase in payoff makes the risk acceptable. To create the table, follow these steps:

Step 1: Enter titles and headings as shown in Figure 3.28. Enter cells B2:G2 as fractional percentages (e.g., 0.06, 0.07, etc.). Then use the **Percentage** format to display the cells as percents.

Step 2: Create the column of year values. First, enter **0** in cell A6 and **10** in A7. Then select the region A6:A7 and drag the Fill Handle down seven more rows.

Step 3: For appearance, center justify the year values as shown in Figure 3.29.

Step 4: Type in **1** for year zero under each interest rate (cells C5:H5).

Step 5: Format region C6:H12 to be in Number format with 4 decimal places.

Step 6: The formula for single-payment compound factor is $[1 + i]^N$, where i is the interest rate and N is the number of interest computations—in this case, years. Enter the following formula for the single-payment compound amount factor in cell C7:

= (1+C$3)^$B7.

Notice that dollar signs have been included to indicate that the interest rate is always in row 3 and the year is always in column B.

Step 7: Use the Fill Handle to copy the formula to cells C8:C12.

Step 8 : Select the region C7:C12. Drag the Fill Handle to the right until the whole table is populated. Your completed table should resemble Figure 3.30.

Notice that the equation for a compound amount factor was only entered once. All that was necessary to complete the rest of the table was to copy the equation in cell C7 to the rest of the cells.

	A	B	C	D	E	F	G	H
1	**Compound Amount Factors**							
2								
3		**Interest Rate**	6%	7%	8%	9%	10%	11%
4								
5		**Year**						

Figure 3.28. Titles and headings.

	A	B	C	D	E	F	G	H
1	**Compound Amount Factors**							
2								
3		**Interest Rate**	6%	7%	8%	9%	10%	11%
4								
5		Year						
6		0	1.0000	1.0000	1.0000	1.0000	1.0000	1.0000
7		10						
8		20						
9		30						
10		40						
11		50						
12		60						

Figure 3.29. Years added and justified and data for year zero entered.

	A	B	C	D	E	F	G	H
1	Compound Amount Factors							
2								
3		Interest Rate	6%	7%	8%	9%	10%	11%
4								
5		Year						
6		0	1.0000	1.0000	1.0000	1.0000	1.0000	1.0000
7		10	1.7908	1.9672	2.1589	2.3674	2.5937	2.8394
8		20	3.2071	3.8697	4.6610	5.6044	6.7275	8.0623
9		30	5.7435	7.6123	10.0627	13.2677	17.4494	22.8923
10		40	10.2857	14.9745	21.7245	31.4094	45.2593	65.0009
11		50	18.4202	29.4570	46.9016	74.3575	117.3909	184.5648
12		60	32.9877	57.9464	101.2571	176.0313	304.4816	524.0572

Figure 3.30. The completed single-payment compound amount factor table.

APPLICATION—INTERACTIVE DC CIRCUIT ANALYZER

A general expression for the current I in a DC transient circuit is

$$I(t) = I_\infty + (I_0 - I_\infty)e^{-t/T},$$

where

I_0 is an initial value at the instant of sudden change,
I_∞ is the current at time $t = \infty$,
$T = RC$ is the time constant for a series R–C circuit, and
$T = L/R$ is the time constant for a series R–L circuit.

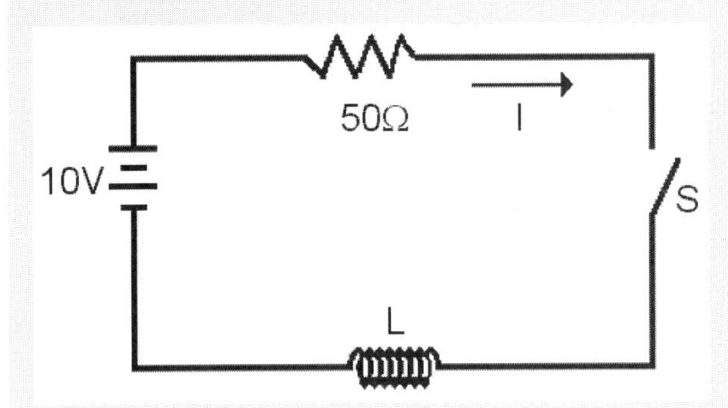

Figure 3.31. DC circuit.

If the switch is closed at $t = 0$, then we can calculate the current in the circuit after three time constants ($t = 3T$). Since $I_0 = 0$ and $I_\infty = V/R$, a worksheet can be set up that calculates $I(3T)$ for various values of V and R. If I_0, V, and R are placed into cells B5, C5, and D5, respectively, then the Excel formula for I_∞ is

```
= C5 / D5 ,
```

and the Excel formula for $I(3T)$ is

```
= E5 + (B5 – E5) * EXP(F5) .
```

If you enter the voltage (10 volts) and resistance (50 ohms) from the circuit displayed in Fig 3.31, then your results should resemble Figure 3.32. The current at three time constants after the switch is closed equals 0.190043 A.

You can use the worksheet in Figure 3.32 to interactively compute the current by entering various values for V and R.

I(0)	V	R	I(infinity)	-t/T	I(3T)
0	10	50	0.2	-3	0.190043

Figure 3.32. Interactive calculator for finding the current in a DC transient circuit.

KEY TERMS

absolute reference	dependents	diagonal
formula auditing	macro	matrix
matrix order	matrix product	precedents
prototype	reference pattern	relative reference
transpose	Visual Basic editor	

SUMMARY

The entry of mathematical formulas and functions is an important use of worksheets for engineers. In this chapter, you were introduced to the use of formulas and functions in Excel. The mathematical and engineering functions were emphasized. You were also shown several methods for debugging and auditing worksheets. Finally, the methods for recording, executing, and editing a macro were demonstrated.

Problems

1. Place the number 10 in cell A1. Create an Excel formula that computes

$$f(x) = x^2 - 4x + 3$$

for cell A1.

2. Place the values for $x = 5$ and $y = 7$ in cells A1 and A2, respectively. Create an Excel formula that computes

$$f(x,y) = y^3 - 10x^2$$

for cells A1 and A2.

3. Place the numbers 1,2,... 10 in cells A1:A10. Create an Excel formula that computes

$$f(x) = \ln x + \sin x$$

for cell A1. Use the Fill Handle and drag the formula over cells A2:A10 to compute

$f(x)$ for 2,3,... 10.

4. For a damped oscillation as depicted in Figure 3.33, the displacement of a structure is defined by the equation

$$f(x) = 8\,e^{-kt}\cos(\omega t),$$

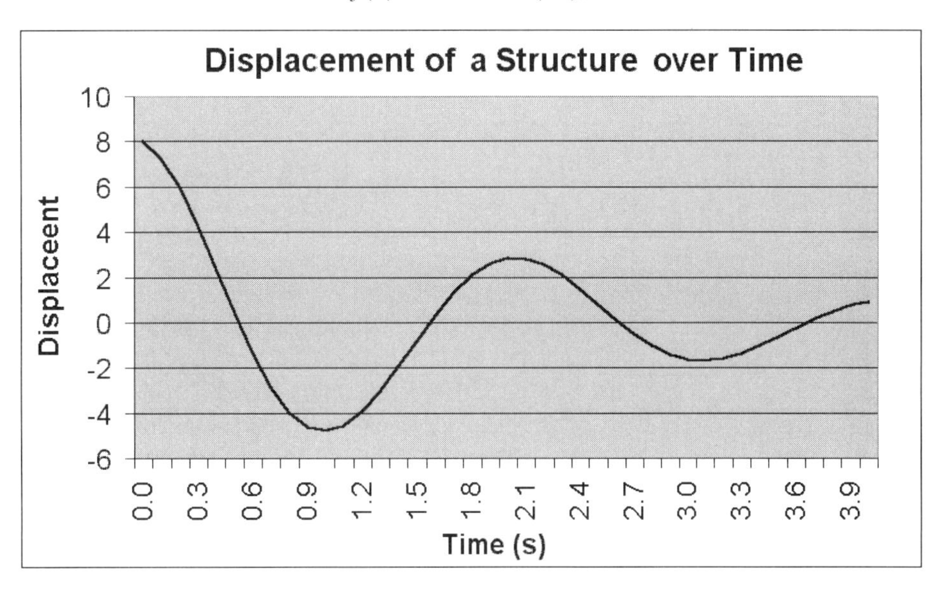

Figure 3.33. Damped oscillation representing displacement of a structure.

where $k = 0.5$ and the frequency $\omega = 3$. Create an Excel formula for this equation. Compute $f(x)$ for $t = 0.0, 0.1, 0.2, \ldots 4.0$ seconds. What is the value of $f(x)$ for $t = 3.6$?

5. The formula that calculates the number of combinations of r objects taken from a collection of n objects is

$$C(n,r) = \frac{n!}{(n - r)!\,r!}.$$

The exclamation point is the mathematical symbol for the factorial operation. The factorial of a number $n = n \times (n - 1) \times (n - 2) \times, \ldots \times 3 \times 2 \times 1$. Thus, the factorial of

$$4 = 4 \times 3 \times 2 \times 1 = 24.$$

The Excel function for factorial is called FACT. The preceding formula can be used to compute the number of ways a committee of 6 people can be chosen from a group of 8 people:

$$C(8,6) = \frac{8!}{(8 - 6)!6!} = 28.$$

Write an Excel equation to calculate combinations. Use it to compute how many 5-card hands may be drawn from a deck of 52 cards.

6. Excel has a number of predefined logical functions. One of these, the IF() function, has the following syntax:

IF (EXP, T, F).

The effect of the function is to evaluate the expression EXP, which must be a logical expression. If the expression is true, then T is returned. If the expression is false, then F is returned. For example, the use of the IF function

IF (X < 200, X, "Cholesterol is too high")

returns the value of X if X is less than 200. But if X is greater than, or equal to, 200, then the text statement "Cholesterol is too high" is placed in the selected cell.

7. Expand the sample of a quadratic equation example in this chapter to test for divide by zero. If the expression $2a = 0$ is true, then display "Divide by Zero"; otherwise, return the value of $2a$.

8. Perform a similar test for $b^2 - 4ac \geq 0$. Display "Requires Complex Number" if the test is false.

9. Neglecting air resistance, the horizontal range of a projectile fired into the air at angle θ degrees is given by the formula

$$R = \frac{2V^2 \sin\theta \cos\theta}{g}.$$

Create a worksheet that computes R for a selected initial velocity V and firing angle θ. Use $g = 9.81$ meters/sec^2. Convert degrees to radians by using the RADIANS() function. To test your results, an initial velocity of 150 meters/sec and firing angle of 25° should result in $R = 1{,}756$ meters.

10. Two other frequently performed matrix operations are the calculation of the *determinant* of a matrix and the *inverse* of a matrix. The Excel functions for these operations are MDETERM and MINVERSE, respectively. Create matrices A and B as shown in Figure 3.20. Compute the matrix inverse and determinant of A and B.

11. Create a macro that computes, labels, and displays the determinant and inverse of a 3×3 matrix which is typed into cells (A1:C3). Create a shortcut key to execute the macro.

4

Working with Charts

4.1 USING THE CHART WIZARD TO CREATE AN XY SCATTER CHART

The Chart Wizard guides you through the construction of a chart. Once the chart is built, its components may be modified. Using the Chart Wizard is probably the easiest way to learn to create a chart in Excel.

Before proceeding, create the worksheet in Figure 4.1. These data were collected by measuring the current (I) in amperes (A) across a resistor for nine measured voltages (V). Ohm's law, which is

$$V = IR,$$

states that the relationship between V and I is linear if temperature is kept relatively constant. In this section, using the Chart Wizard, we will create an XY scatter plot of the data in Figure 4.1. The scatter plot is useful for visualizing relationships among data.

SECTION

OBJECTIVES

After reading this chapter, you should be able to:

- Create a chart using the Chart Wizard.
- Create a chart using shortcut keys.
- Create line charts and XY scatter plots from worksheet data.
- Preview and print a chart.
- Format chart legends, axes, and titles.
- Scale axes and create error bars.

Complete the following steps to create a scatter plot of the data in Figure 4.1:

Step 1: Select the region containing the data (A2:B10).

Step 2: Start the Chart Wizard by choosing the Chart Wizard icon from the Standard toolbar. You can also choose **Insert → Chart** from the Menu bar. The first Chart Wizard dialog box will appear as shown in Figure 4.2. This Chart Wizard dialog box prompts you to choose a chart type.

	A	B
1	Potential (V)	Current (A)
2	6.970	0.051
3	5.960	0.044
4	4.950	0.038
5	3.980	0.032
6	3.030	0.025
7	1.910	0.018
8	1.020	0.012
9	0.500	0.008
10	0.200	0.001

Figure 4.1. Data collected by measuring current across a 150 Ω resistor.

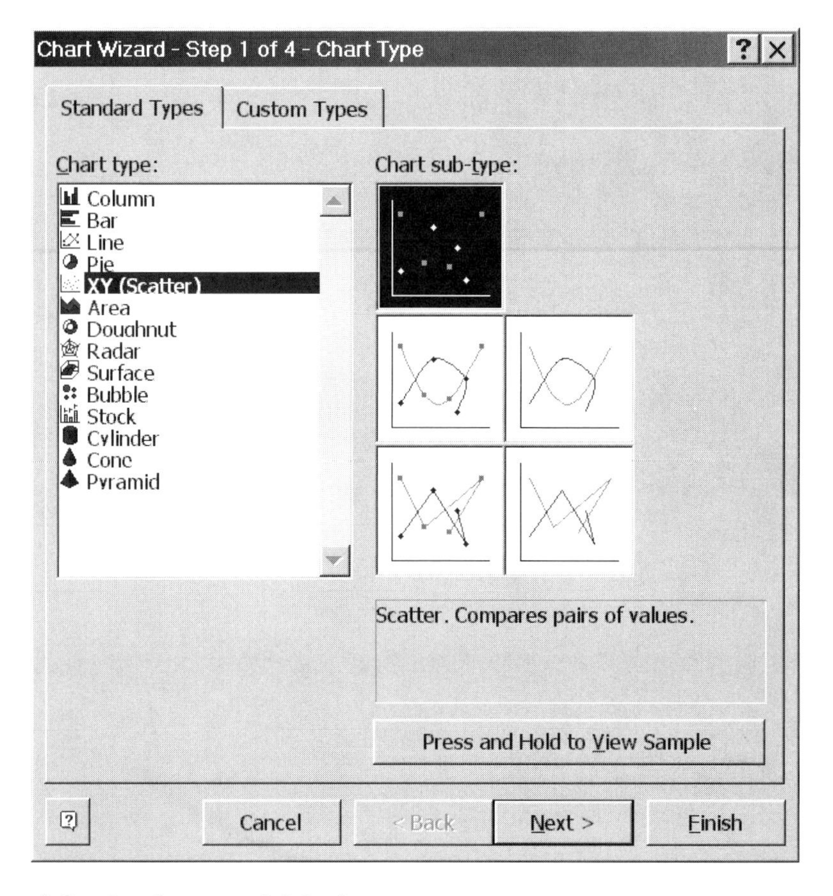

Figure 4.2. The Chart Wizard dialog box, Step 1.

Step 3: Select **XY (Scatter)** from the list labeled *Chart type*. Select the top-left box from the area labeled *Chart sub-type* and then choose the **Next** button to proceed. The second Chart Wizard dialog box will now appear as shown in Figure 4.3.

Step 4: Choose the **Series** tab. Excel has chosen, by default, to plot cells (A2:A10) on the X axis and cells (B2:B10) on the Y axis.

Step 5: Choose the small chart icon on the right end of the boxes labeled *X Values* or *Y Values* to modify the selected regions for the X values or the Y values. The Source Data dialog box will appear as depicted in Figure 4.4. The currently selected region will be surrounded by a dashed line. If you wish to change the X values, use the mouse to select a region and then choose the small chart icon on the right end of the Source Data dialog box to return to the Chart Wizard. For our current example, the X values do not need to be modified. Choose **Next** to proceed.

Step 6: The third Chart Wizard dialog box should appear as depicted in Figure 4.5. This dialog box guides you through the chart formatting options. From this dialog box you can create and modify the chart titles, axes, gridlines, data labels, and the legend.

Step 7: Choose the **Titles** tab and then type a chart title and X and Y axis titles, as depicted in Figure 4.5.

Step 8: Remove the legend by choosing the **Legend** tab. Make sure that the box titled *Show Legend* is not checked.

Step 9: Choose **Next** to proceed to the fourth and final step.

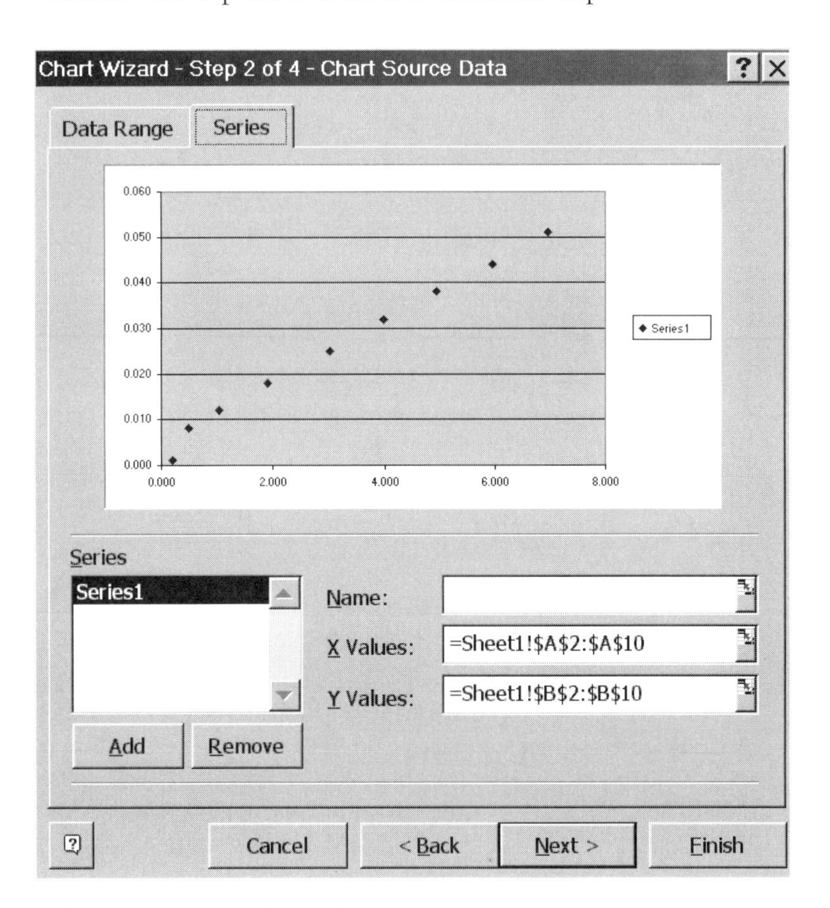

Figure 4.3. The Chart Wizard dialog box, Step 2.

	A	B
1	Potential (V)	Current (A)
2	6.970	0.051
3	5.960	0.044
4	4.950	0.038
5	3.980	0.032
6	3.030	0.025
7	1.910	0.018
8	1.020	0.012
9	0.500	0.008
10	0.200	0.001
11		

Chart Wizard - Step 2 of 4 - Chart Source Data - X Values:

=Sheet1!$A$2:$A$10

Figure 4.4. Choosing a region for the X values.

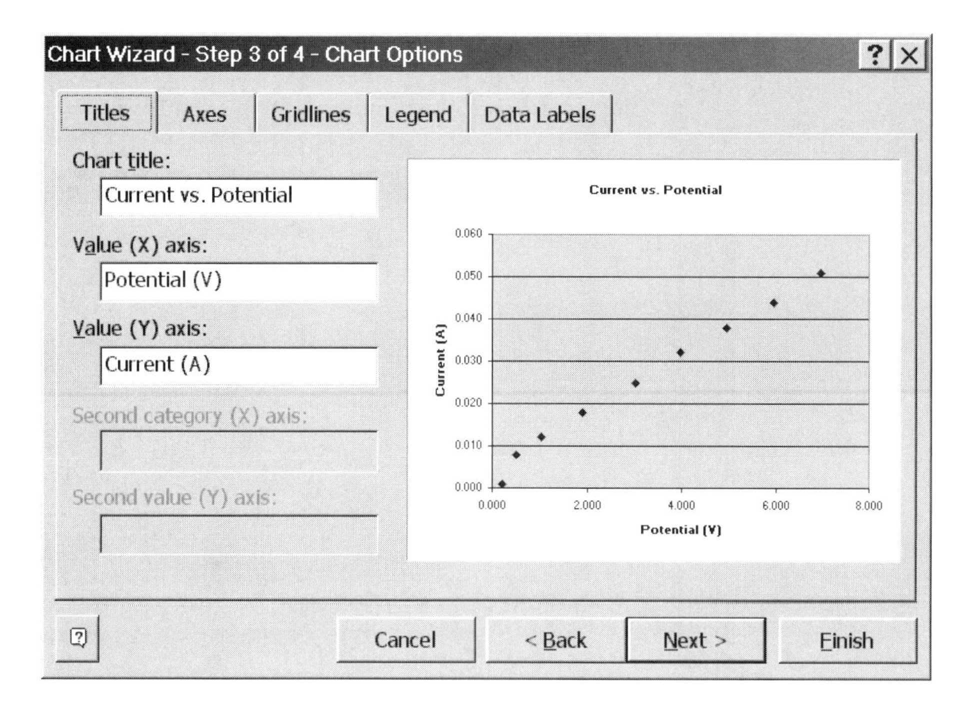

Figure 4.5. The Chart Wizard dialog box, Step 3.

Step 10: The fourth Chart Wizard dialog box should now appear, as depicted in Figure 4.6. This dialog box prompts you to choose a location for the chart. There are two choices.

The first choice, titled *As new sheet*, will place the chart as a separate worksheet in the current workbook. If a name is not typed, then Excel provides a default name (e.g., Chart 1, Chart 2, etc.). The second choice, titled *As object in*, will place the chart as an object in the selected worksheet.

Step 11: Choose **As new sheet** and then click the **Finish** button. The completed XY Scatter chart will appear as a new worksheet, as shown in Figure 4.7.

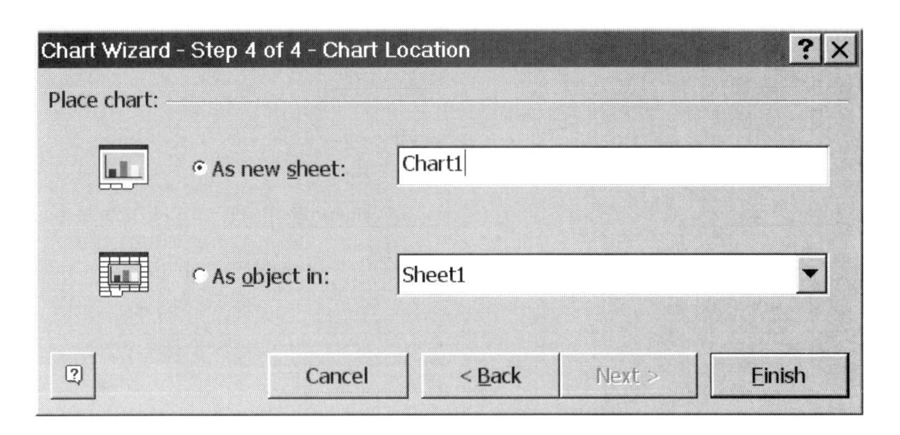

Figure 4.6. The Chart Wizard dialog box, Step 4.

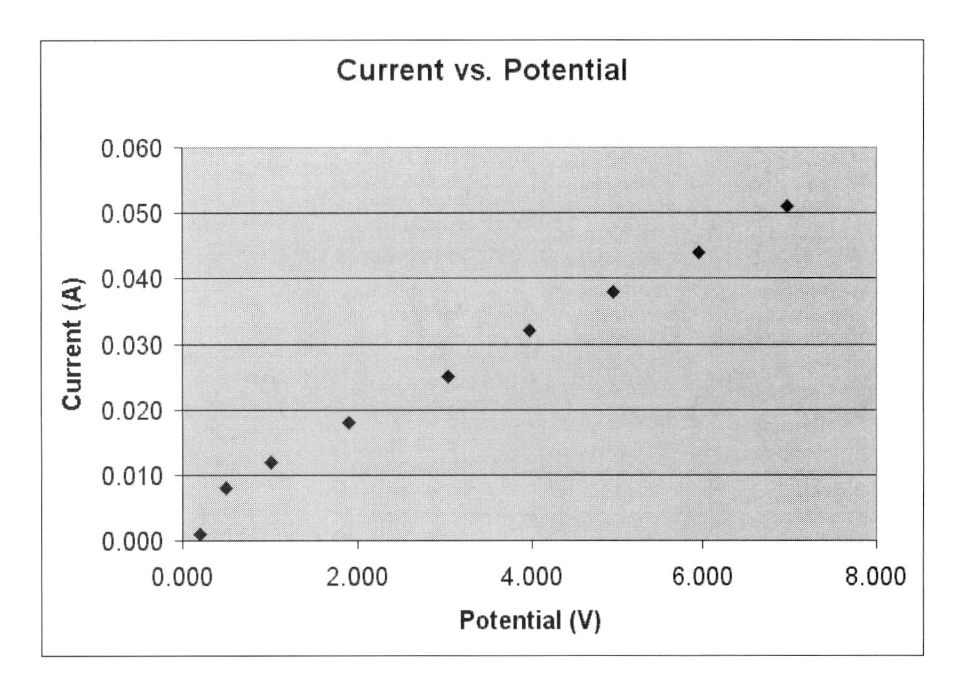

Figure 4.7. The completed XY Scatter chart.

4.2 CREATING A CHART BY USING SHORTCUT KEYS

If you frequently create charts of the same type, then you can use a shortcut to quickly create a chart. The **F11** shortcut key creates a chart and formats it in the default chart type.

To set the default chart type,

Step 1: Choose **Chart** → **Chart Type** from the Menu bar.

Step 2: The Chart Type dialog box will appear.

Step 3: Choose a chart type and subtype from the dialog box, then click **Set as default chart**. A confirm box will appear, as shown in Figure 4.8. Choose **Yes**. You will be returned to the Chart Type dialog box.

Step 4: Choose **OK** to exit the Chart Type dialog box.

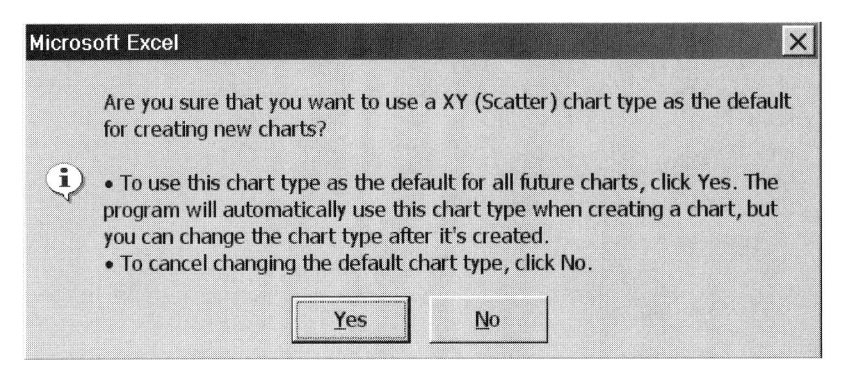
Microsoft Excel

Are you sure that you want to use a XY (Scatter) chart type as the default for creating new charts?

• To use this chart type as the default for all future charts, click Yes. The program will automatically use this chart type when creating a chart, but you can change the chart type after it's created.
• To cancel changing the default chart type, click No.

Yes | No

Figure 4.8. Box to confirm the default chart type.

PRACTICE!

To practice creating a chart by using the **F11** shortcut key,

Step 1: Set the default chart type to **XY scatter**, as described earlier.

Step 2: Select the region A1:B10 from Figure 4.1. Note that we have included the column headings in the region.

Step 3: Press the **F11** key. A chart will appear, as shown in Figure 4.9. Note that the chart title and labels could use improvement. Fortunately, chart elements can easily be modified. In Section 4.5, you will be shown how to modify and format chart elements.

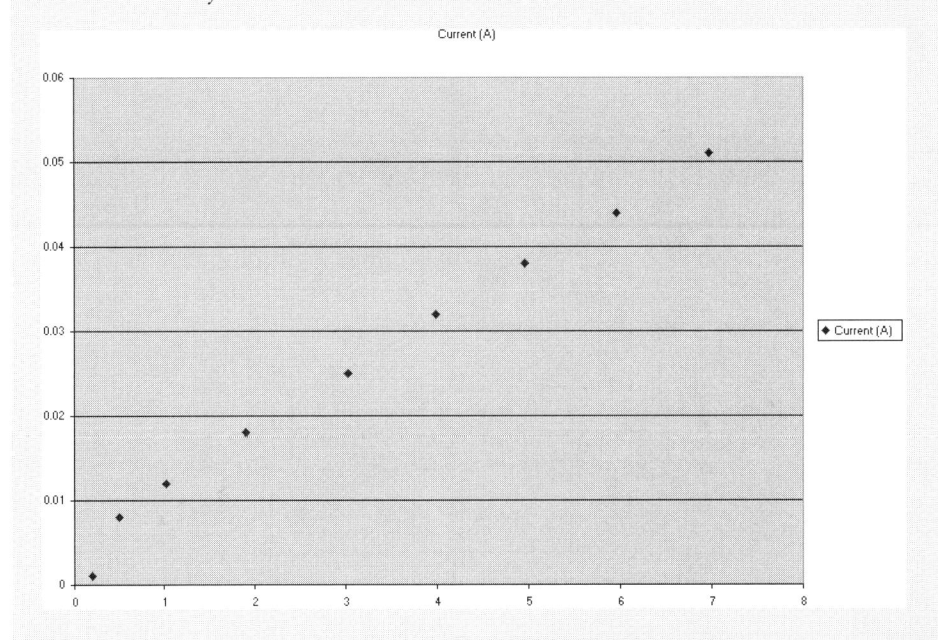

Figure 4.9. Creating a chart with the F11 shortcut key.

Excel makes several assumptions when automatically creating a chart. If the data do not follow Excel's conventions, then surprising results can sometimes occur. If the chart does not turn out as expected, the chart can always be reformatted, as described in the next several sections.

Some of the assumptions made by Excel are as follows:

- Excel orients the chart so that the data for the X category are taken from the longest side of the selected region. In our example (from Figure 4.1), the longest side of the selected region is vertical.
- If the contents of the cells along the short side of the selected region contain text, then they are used as labels for the data series in the legend. In our example, the label *Current (A)* is taken from the top cell of the short (horizontal) side of the selected region. If the cells contain numbers, then the default data series names are used (Series 1, Series 2, etc.).
- If the contents of the cells along the long side of the selected region contain text, then they are used as X category labels. If the cells contain numbers, then Excel assumes that the cells contain a data series.

4.3 PREVIEWING AND PRINTING CHARTS

To preview a chart before printing, choose **File → Print Preview** from the Menu bar. You can also choose the **Print Preview** icon from the Standard toolbar.

A chart that is embedded within a worksheet will print with the worksheet by default. If you select the chart before choosing print preview, then the embedded chart can be printed separately. A chart that is formatted as a separate worksheet will, by default, be printed separately.

For example,

Step 1: Select the worksheet containing your completed XY Scatter chart from Figure 4.7.

Step 2: Choose **File → Print Preview** from the Menu bar. The Print Preview dialog box will appear, as shown in Figure 4.10.

Step 3: Select the **Margins** button, and the margin lines will appear as depicted in Figure 4.10. The margin lines can be dragged to resize and reshape the chart. The **Zoom** button can be used to focus on chart detail.

Step 4: Choose the **Setup** button to select more print formatting options. The Page Setup dialog box will appear as depicted in Figure 4.11. From the tabs on this dialog box, you may select portrait or landscape mode, select the chart size, select the printing quality, manually specify margins, and insert headers or footers.

Step 5: Choose **OK** to exit the Page Setup dialog box, when you are satisfied with your selections. Then choose **Print**.

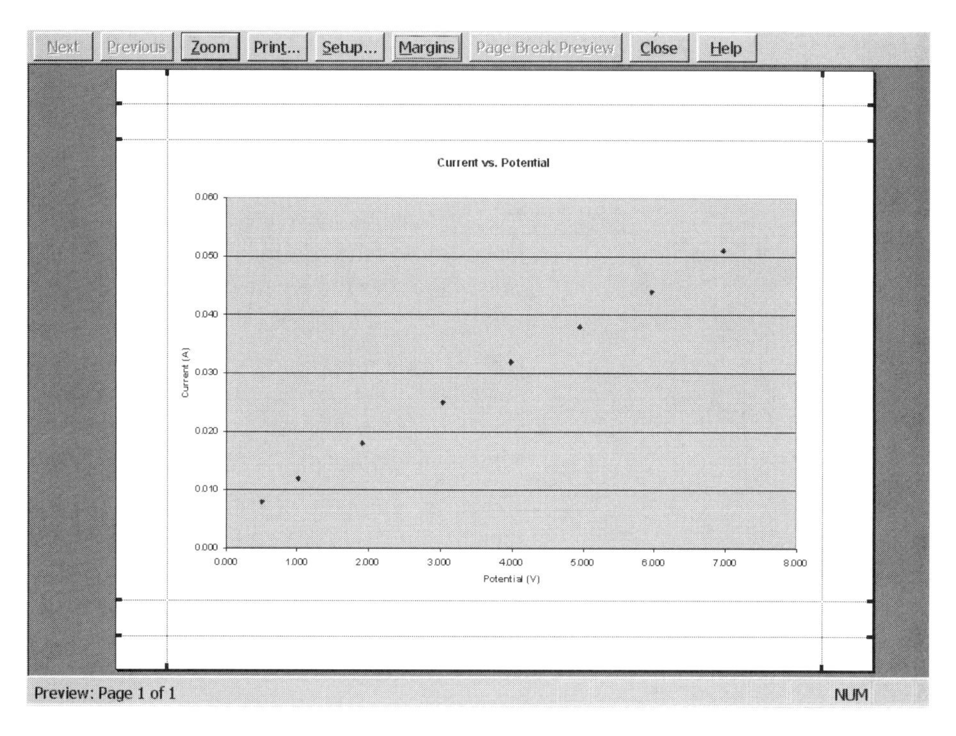

Figure 4.10. The Print Preview dialog box.

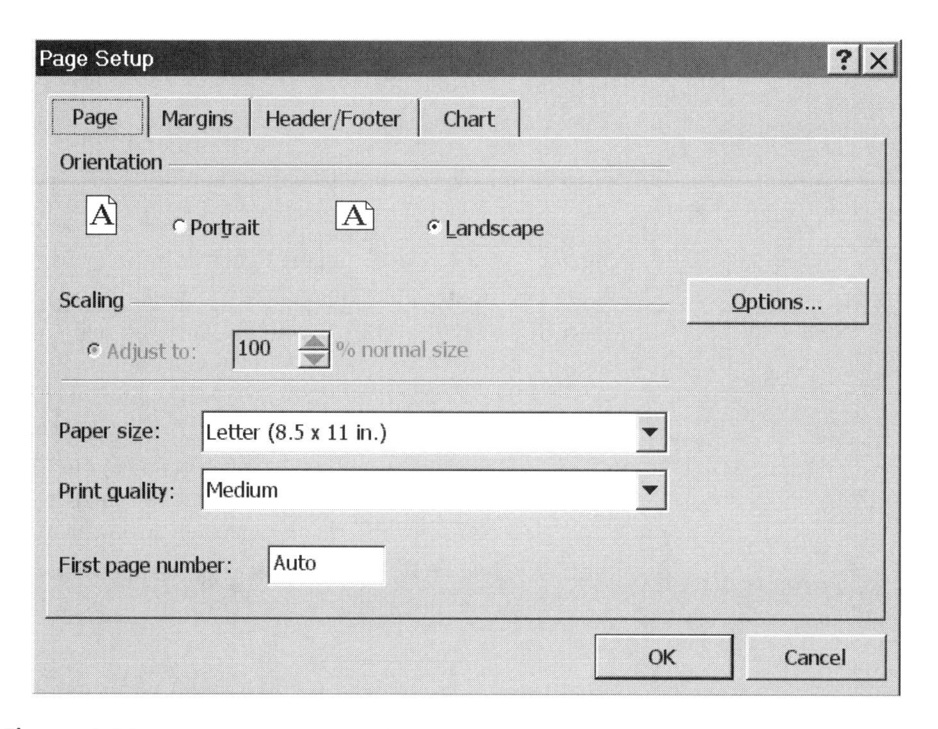

Figure 4.11. The Page Setup dialog box.

4.4 ADDING AND EDITING CHART DATA

4.4.1 Adding Data Points

Data can be added or removed from a chart in the following manner:

Step 1: Activate the chart by clicking on it.

Step 2: Choose **Chart → Add Data** from the Menu bar. The Add Data dialog box will appear, as shown in Figure 4.12. Choose the small table box near the right side of the window labeled *Range*.

Step 3: Select a new region to add to the chart. In Figure 4.13, we have selected another data point to add to the chart's source data. After selecting the new data range, the Paste Special dialog box will appear. (See Figure 4.14.)

Step 4: Check the box labeled *New point(s)*, since we want to add a new data point to the existing data series.

Step 5: Check the box labeled *Columns*, because, in our example, the Y values (Current) are listed in columns.

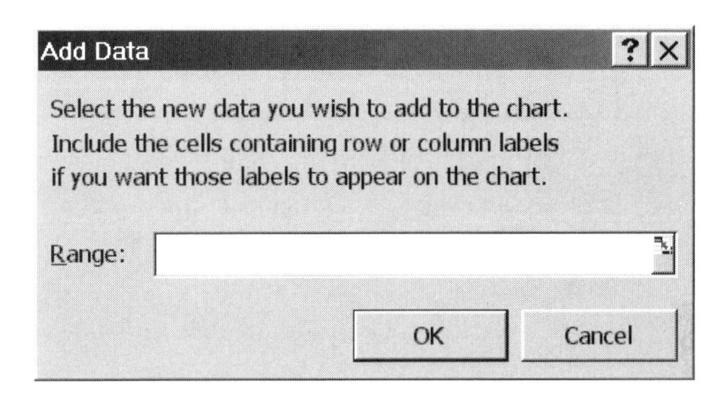

Figure 4.12. The Add Data dialog box.

	A	B
1	Potential (V)	Current (A)
2	6.970	0.051
3	5.960	0.044
4	4.950	0.038
5	3.980	0.032
6	3.030	0.025
7	1.910	0.018
8	1.020	0.012
9	0.500	0.008
10	0.200	0.001
11	0.150	0.001
12		
13		
14		

Add Data - Range:
=Sheet1!$A$11:$B$11

Figure 4.13. Adding data to a chart.

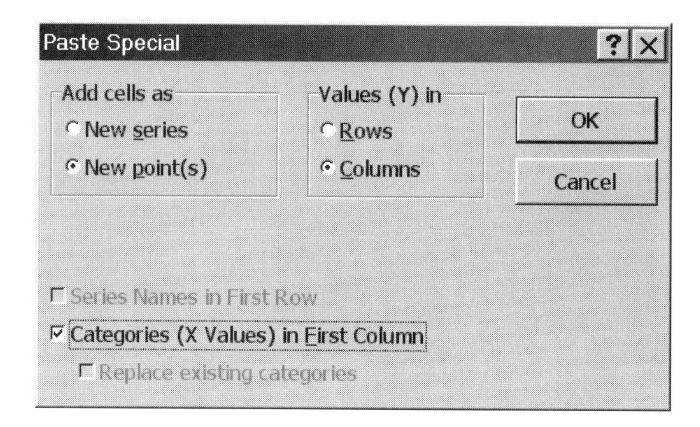

Figure 4.14. The Paste Special dialog box.

Step 6: Check the box labeled *Categories (X Values) in First Column*, since our X values (Potential) are listed in the first column.

Step 7: Click **OK**, and the new data point will be added to the chart.

Another method for adding (or deleting) data points is to modify the source data definition as follows:

Step 1: Activate the chart by clicking once on the chart.

Step 2: Choose **Chart → Source Data** from the Menu bar.

Step 3: The Source Data dialog box will appear, as depicted in Figure 4.15.

Step 4: Choose the **Data Range** tab and modify the contents of the box labeled *Data range* to add, modify, or delete source data points.

4.4.2 Multiple Data Series

In the examples of charts so far, we have used a single data series. A *data series* is a collection of related data points that are to be represented as a unit. For example, the points in a data series are connected by a single line in a line chart. The data for a series would likely be represented by a separate row or column in the worksheet.

Table 4.1 shows data representing the low-flow rates for two tributaries of the Pecos River. The lowest one-day flow rate per year (in cubic feet per second) is shown. Create a worksheet that contains the data in Table 4.1 and use it for the rest of the examples in this chapter.

In the following example, we'll show how to create a chart with two data series:

Step 1: Once you have created a worksheet containing the data and titles in Table 4.1 select the region containing the data and the column headings.

Step 2: Choose **Insert → Chart** from the Menu bar.

Step 3: Follow the Chart Wizard instructions. Select **Line Chart** for a chart type and select the first line chart subtype. The second Chart Wizard box will appear.

Step 4: Select the **Series** tab. The results should resemble Figure 4.16. The Chart Wizard has created three data series, one for each column in the selected region of the worksheet.

Step 5: Remove the series titled *Year* from the Series box by selecting **Year** and clicking the **Remove** button to enable you to make the first column (Year) to be the X-axis data label.

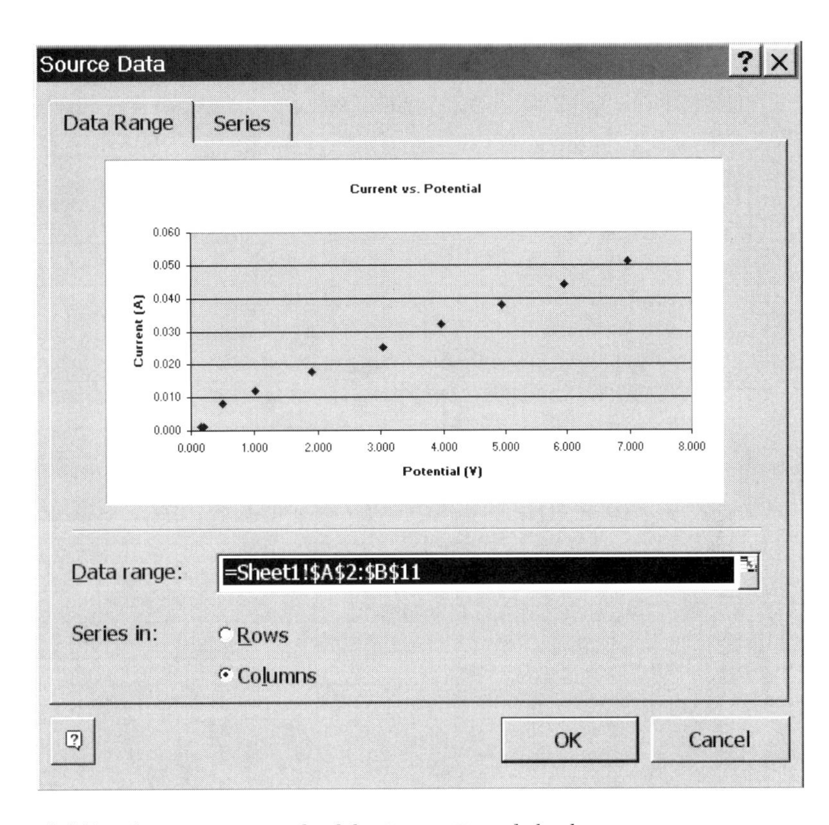

Figure 4.15. The Data Range tab of the Source Data dialog box.

TABLE 4.1 Annual Low-Flow Rate of Pecos River

Year	East Branch Flow (cfs)	West Branch Flow (cfs)
87	221	222
88	354	315
89	200	175
90	373	400
91	248	204
92	323	325
93	216	188
94	195	202
95	266	254
96	182	176

Step 6: Add the range for the Year column to the box titled *Category (X) axis labels*. You can also choose the small table icon on the right side of the box labeled *Category (X) axis labels* and select a range with the mouse.

Step 7: Complete the Chart Wizard. The resulting chart (without chart titles or axis titles) is depicted in Figure 4.17. Save this chart and use it as a sample in the next section on chart formatting.

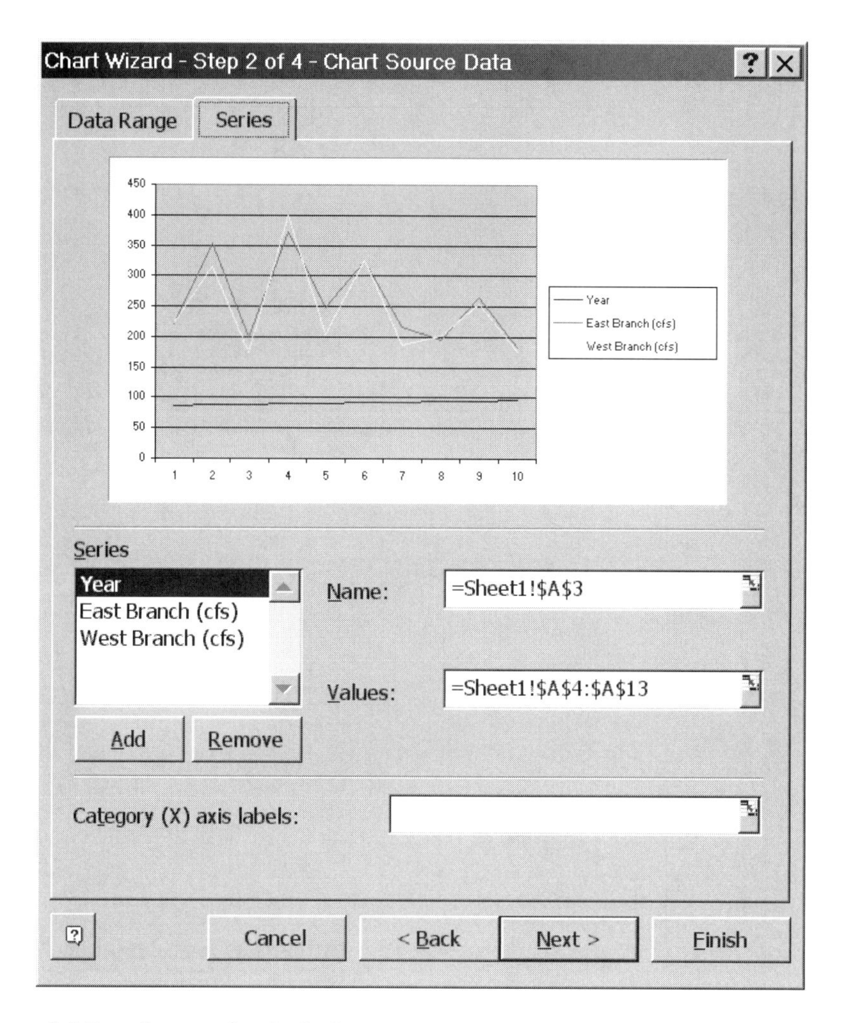

Figure 4.16. Selection of multiple data series.

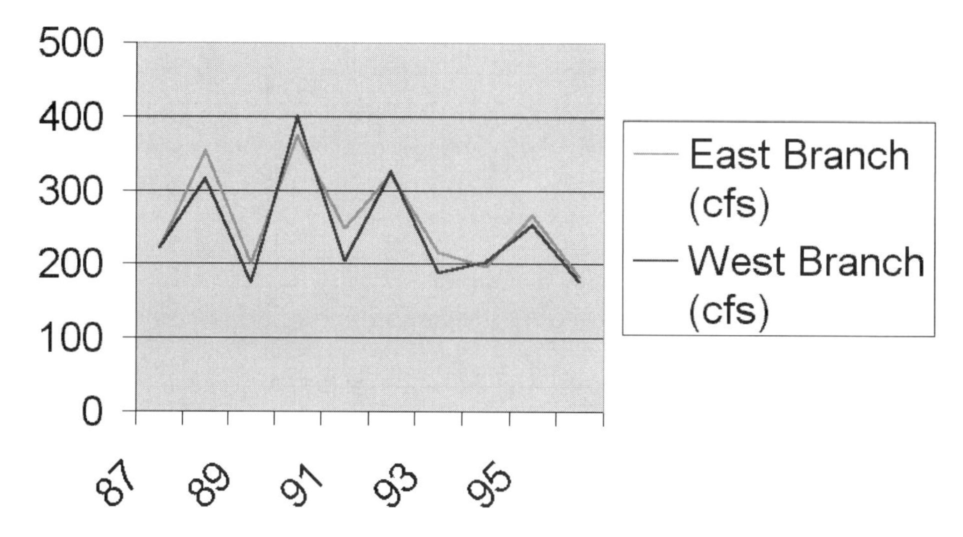

Figure 4.17. Example of a line chart with two data series.

4.5 FORMATTING CHARTS

4.5.1 Creating Chart Objects

A chart consists of a number of elements called *chart objects*. Examples of chart objects are the chart *legend*,[1] the chart title, the *X and Y axes*,[2] and the plot area. Carefully chosen titles, labels, and legends and appropriate axes will increase the quality of your chart.

Each chart object can be separately formatted and customized. It is usually faster to initially create a chart object with the default formatting. This will give you the general look of the chart. Then you can customize each object separately.

To create and enter data into chart objects,

Step 1: Open a worksheet that contains a chart.

Step 2: Click once on the chart to select it. The Menu bar will change, and a Chart menu item will appear on the menu.

Step 3: Choose **Chart → Chart Options** from the Menu bar. The Chart Options dialog box will appear, as shown in Figure 4.18. The Chart Options dialog box contains six tabs that are used, respectively, for entering and formatting the titles, axes, gridlines, legend, data labels, and data table.

Step 4: Select the **Titles** tab. Type an appropriate chart title and labels for the X and Y axes:

- Title: Annual Low Flows for Pecos River
- X axis: Year
- Y axis: Lowest Flow Day (cfs)

Step 5: Select the **Gridlines** tab and check the box labeled *Value (Y) Axis Major Gridlines*. Uncheck all other boxes on the screen.

Step 6: Select the **Legend** tab and check the box labeled *Show legend*.

Step 7: Click **OK**, when you are satisfied with your chart options. Your chart should resemble Figure 4.19.

4.5.2 Formatting Individual Chart Objects

Once created, chart objects can be individually formatted.

To format an individual object,

Step 1: Select the object with the left mouse button. When selected, an object is highlighted and surrounded by a box. If the mouse cursor is positioned over the object, then the name of the object is displayed in a Tool Tip.

Step 2: Select the legend object in your chart. It should resemble Figure 4.20.

Step 3: A selected object can be moved by dragging the object. Try dragging the legend on your chart to a different location.

Step 4: A selected object can be resized by dragging one of the edit points (the small black boxes) on the selected object. Try resizing the legend on your chart.

Step 5: Click the right mouse button and a drop-down menu will appear. The items on the menu will be different for each object. Since you have selected the legend, the menu will have an item titled *Format Legend*. Choose Format Legend from the drop-down menu. The Format Legend dialog box will appear, as depicted in Figure 4.21.

---

[1]A caption or an explanatorey list of symbols used in a chart.

[2]Reference lines in a coordinated system.

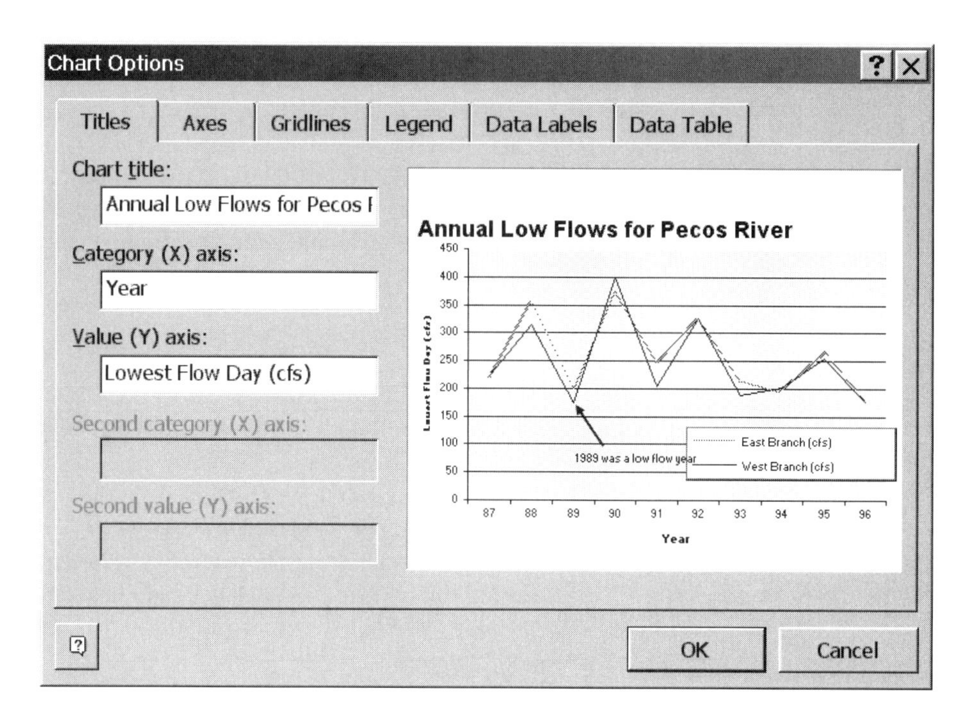

Figure 4.18. The Chart Options dialog box.

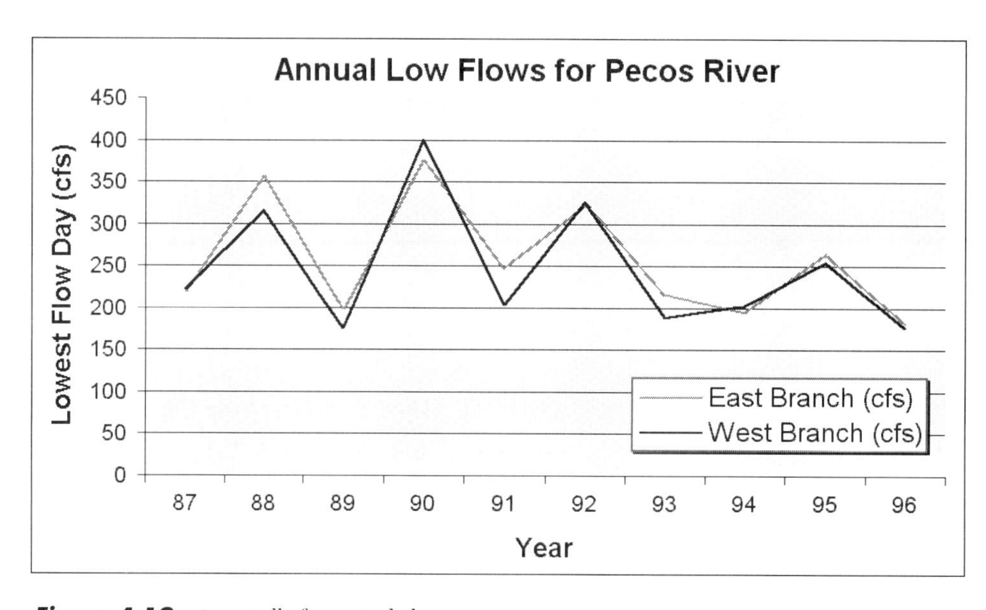

Figure 4.19. A partially formatted chart.

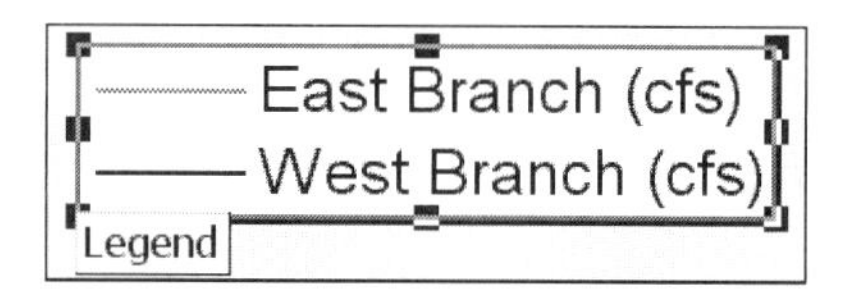

Figure 4.20. A selected chart object (the legend).

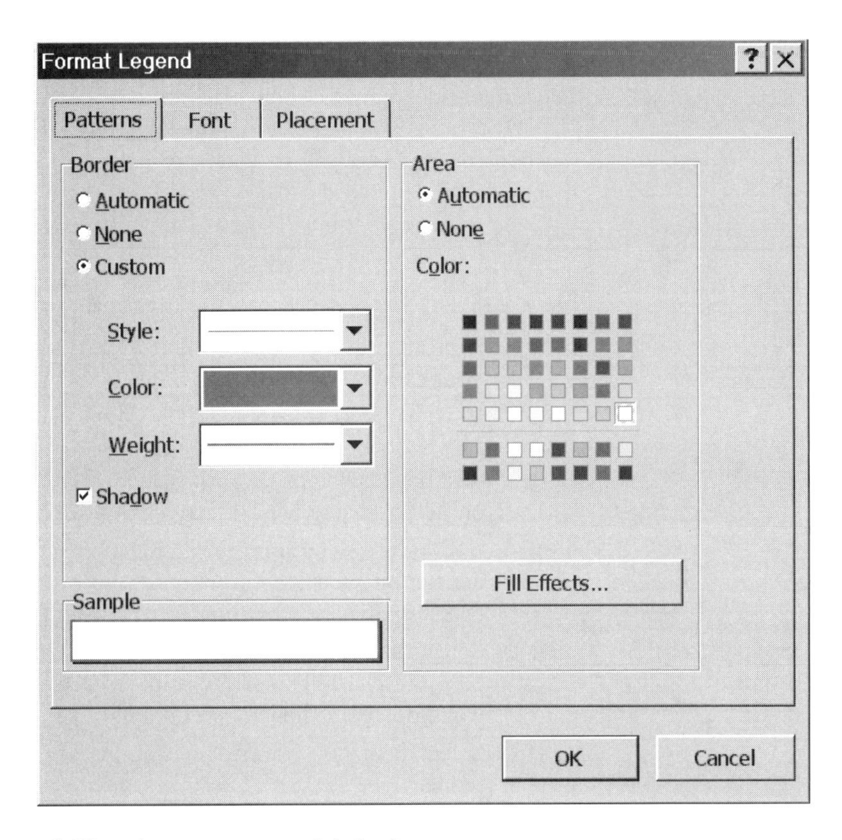

Figure 4.21. The Format Legend dialog box.

Step 6: Each chart object has a similar formatting dialog box. From the formatting dialog boxes, you can modify the border patterns, colors, font characteristics, and object placement.

PROFESSIONAL SUCCESS—FORMATTING CHARTS

The appearance of a chart in a document is important. A chart can make a lasting, visual impression that summarizes or exemplifies the main points of your presentation or document. The following formatting guidelines will help you create a professional-looking chart:

- A chart title should contain a clear, concise description of the chart contents.
- Create a label for each axis that contains, at a minimum, the name of the variable and the units of measurement that were used.
- Create a label for each data series. The labels can be consolidated in a legend if each data series is represented by a distinct color or texture.
- Scale graduations should be included for each axis. The graduation marks may take the form of *gridlines* (uniformly spaced horizontal and vertical lines) or tick marks. The choice of scale graduations can be manipulated somewhat by selecting the **Scale** tab from the Format Axis dialog box.
- Ideally, scale graduations should follow the 1, 2, 5, rule. The *1, 2, 5 rule* states that one should select scale graduations so that the smallest division of the axis is a positive or negative integer power of 10 times 1, 2, or 5. For example, a scale graduation of 0.33 does not follow the rule.

4.5.3 Annotating a Chart

The text for titles and axis labels can be added or modified by selecting and formatting chart objects. At times, you may wish to annotate a chart to emphasize a feature of the chart. The functions on the Drawing toolbar may be used to create arrows or other free-floating shapes. Free-floating text can also be added to highlight or explain a specific data point.

As an example, we will add text and an arrow to the chart in Figure 4.19 to emphasize that 1989 was a low-flow year for both river branches. To create the free-floating text,

Step 1: Select any nontext object in the chart (e.g., the outside border).

Step 2: Type the text, then press **Enter**. For example, type **1989 was a low flow year**. The text should appear inside a small gray box called a Text box.

Step 3: **Format → Text Box** from the Menu bar to modify the font. To increase the font size, the text box must be large enough to hold the new font. If necessary, the Text box may be moved or resized with the mouse. You can edit the text directly on the chart simply by selecting the text box and then typing.

Step 4: Add an arrow by using the Drawing toolbar. If the Drawing toolbar does not appear on your screen, choose **View → Toolbars** from the Menu bar and check the box titled *Drawing*.

Step 5: Select the **Arrow** icon from the Drawing toolbar. Drag and drop an arrow onto the chart. The arrow won't appear yet. However, you will see two small white dots that show the endpoints of the arrow.

Step 6: Select an arrow style by choosing the **Arrow Style** icon from the Drawing toolbar.

Step 7: Select an arrow color by choosing the **Line Color** icon from the Drawing toolbar.

Step 8: Select the thickness of the arrow by choosing the **Line Style** icon from the Drawing toolbar.

Step 9: Move and resize the arrow by selecting the arrow object and dragging the endpoints to the desired location. When finished, click somewhere outside of the chart. Your finished chart should resemble Figure 4.22.

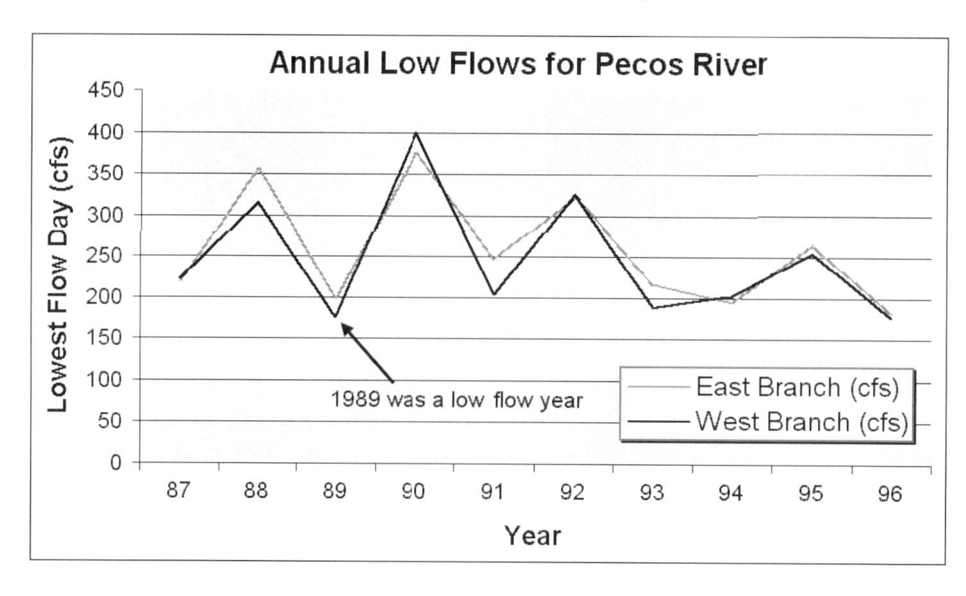

Figure 4.22. Formatted chart with annotation.

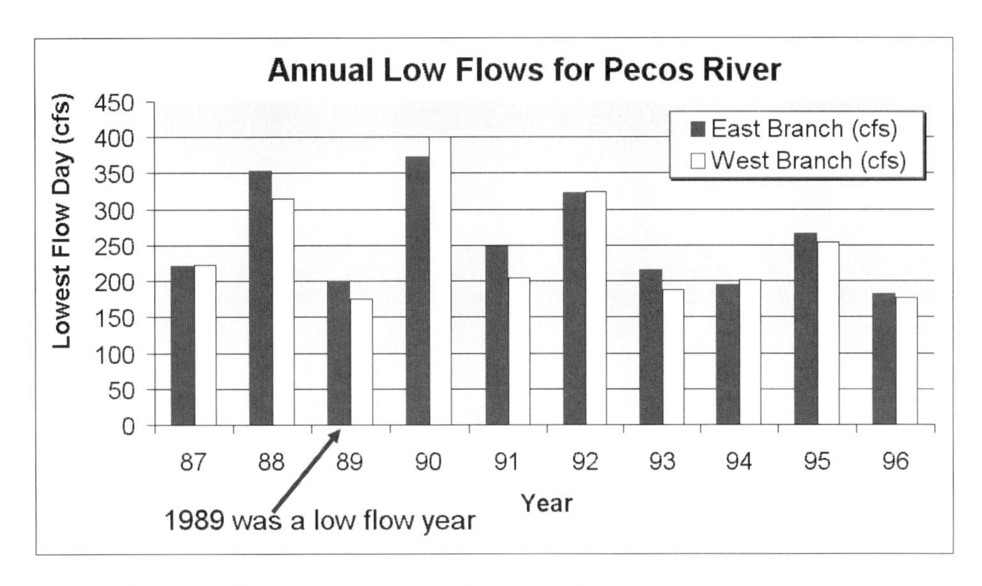

Figure 4.23. Sample column chart, using data from Table 4.1.

4.5.4 Changing Chart Types

The type of chart displayed can be changed after a chart has been created. The same data are used for the new chart type. Not all types of charts are appropriate for some data sets. For example, a pie chart is not appropriate for the river-flow data used in this chapter.

A column chart is an acceptable chart type for the data in Table 4.1. To change the chart type to a column chart,

Step 1: Select your chart.

Step 2: Choose **Chart → Chart Type** from the Menu bar. The Chart Type dialog box will appear.

Step 3: Select **Column** from the box labeled *Chart Type* and the first style from the box labeled *Chart sub-type*. The resulting chart should resemble Figure 4.23. Note that the chart in your worksheet has been replaced. The formatting has been copied over to the new chart. You may have to resize and move some of the chart objects.

4.6 CHARTING FEATURES USEFUL TO ENGINEERS

Excel has many advanced features for formatting charts. Some of the features are particularly useful for engineering applications. These include trend lines, error bars, what-if analysis, axis scaling, and the use of secondary axes.

4.6.1 Scaling an Axis

The scale of an axis is used to delimit the range of the axis as well as the intervals between axis markers. Axis markers are called *ticks*. Large tick marks are specified in *major units*, and small tick marks are specified in *minor units*.

To change the axis formatting,

Step 1: Select the **Value (Y) axis** from your chart of the river-flow data.

Step 2: Choose **Format → Selected Axis** from the Menu bar.

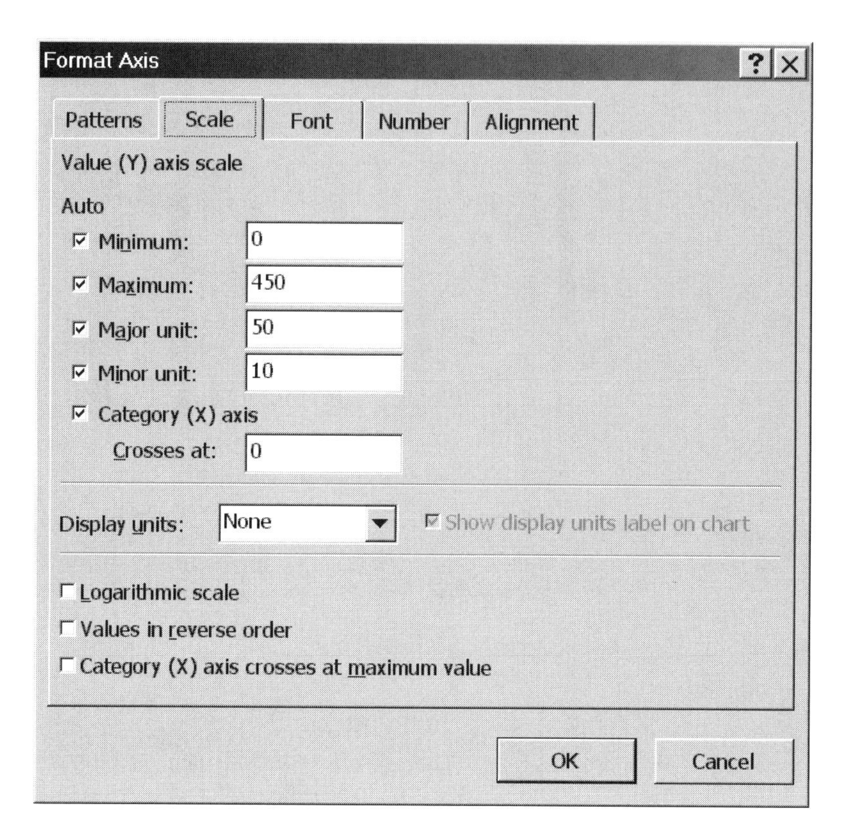

Figure 4.24. The Scale tab of Format Axis dialog box.

Step 3: The Format Axis dialog box will appear, as shown in Figure 4.24. Select the **Scale** tab.

Step 4: From the Scale tab, you can select the minimum and maximum values of the axis, as well as the size of the major and minor tick marks. If the boxes under the *Auto* label are checked, Excel will try to determine the best values for the axis.

Step 5: Of particular interest to engineers is the logarithmic scale. If you select this box, the values of the boxes labeled *Minimum, Maximum, Major unit*, and *Minor unit* will be recalculated to be powers of 10. A logarithmic scale cannot contain values that are less than, or equal to, zero. Illegal values will result in an error message.

Step 6: Choose the **Patterns** tab of the Format Axis dialog box. The Patterns screen is shown in Figure 4.25.

Step 7: From the Patterns screen, you can select the placement of the major and minor tick marks and the placement of the tick-mark labels.

4.6.2 Error Bars

Error bars represent the range of measured or statistical error in a data series. Error bars should not be used unless you understand their purpose.

We will use error bars to indicate that the low-flow measurements in our chart have up to a 5% measurement error in either direction. To add error bars to a data series,

Step 1: Select the series by clicking on them once with the left mouse button. The series will be highlighted and the data points will be emphasized with small black boxes at each point.

Format Axis
Patterns
Scale
Font
Number
Alignment
Lines
Automatic
None
Custom
Style:
Color:
Automatic
Weight:
Sample
Major tick mark type
None
Outside
Inside
Cross
Minor tick mark type
None
Outside
Inside
Cross
Tick mark labels
None
High
Low
Next to axis
OK
Cancel

Figure 4.25. The Patterns tab of the Format Axis dialog box.

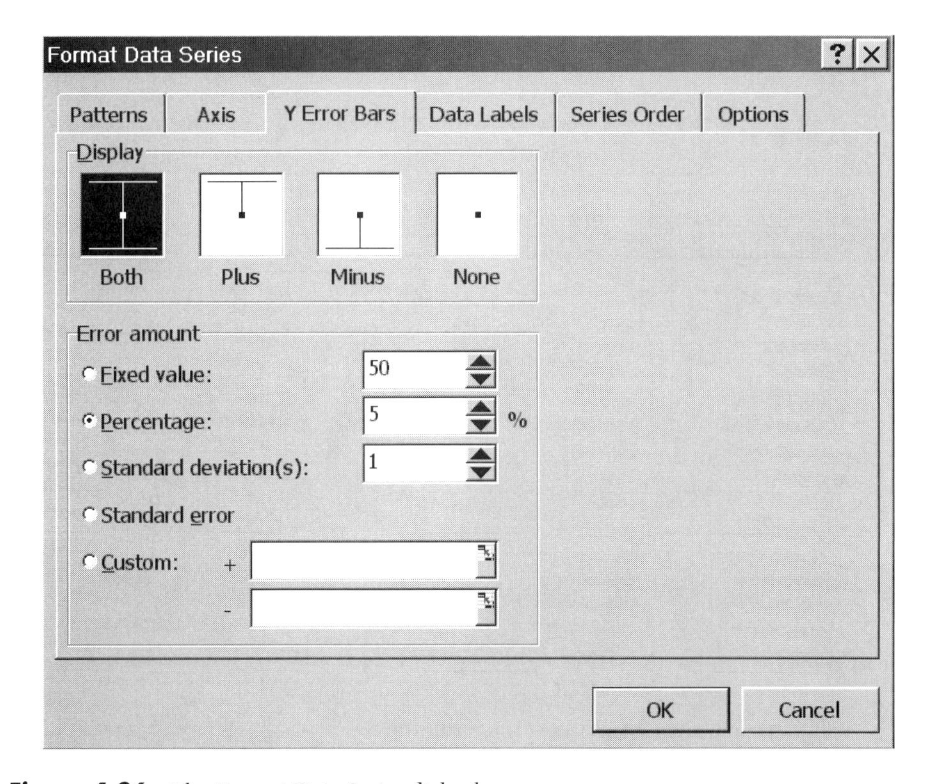

Figure 4.26. The Format Data Series dialog box.

Step 2: Choose **Format** → **Selected Data Series** from the Menu bar. The Format Data Series dialog box will appear.

Step 3: Select the **Y Error Bars** tab. The Format Data Series dialog box should appear, as shown in Figure 4.26.

Step 4: Check the box labeled *Both*.

Step 5: Select the box labeled **Percentage** and type **5** into the Percentage box.

Step 6: Click **OK**. Your chart will now contain error bars around each data point.

APPLICATION—GRAPHING TO EVALUATE A FUNCTION

Visualization can be a big help in trying to understand what your data means, or how a function works. Excel can help with this by allowing you to quickly and easily graph data by evaluating a function. For example, exponentials and hyperbolic sines are commonly used functions for solving differential equations. When solving these equations, you select the appropriate function based on its characteristics. Being able to see a graph of a function is a big help in understanding how the function behaves. Figure 4.27 shows a plot of the hyperbolic sine function from −10 to 10.

HOW'D YOU DO THAT?

To help visualize a function, first evaluate the function over a range of values in a worksheet. Then,

Step 1: Create a new worksheet. Type the label **X** in cell A1. Type the label **sinh(X)** in cell A2.

Step 2: In cells (A2:A22), create the integers from −10 to 10 in increments of one. Use the Fill Handle to simplify this task.

Step 3: Type the formula **=sinh(A2)** in cell B2.

Step 4: Use the Fill Handle to drag the formula through cells (B3:B22)

Step 5: Format cells (B2:B22) as Number with zero decimal places. Your worksheet should look like Figure 4.28.

To plot the computed values,

Step 1: Select the two columns of values, including the headings.

Step 2: Click the Chart Wizard icon on the Standard toolbar.

Step 3: Select an **XY (Scatter)** chart type, and choose the subtype with smoothed lines and no markers. Click **Next** to continue to Step 2.

Step 4: A preview of the graph is displayed. Step 2 of the Chart Wizard provides an opportunity to make changes to the displayed data. Since you don't need to do that, click **Next** to continue to Step 3.

Step 5: Choose the **Titles** tab and add titles and axis labels as follows:

- Title: Plot of hyperbolic sine
- X axis: X
- Y axis: sinh(X)

Step 6: Choose the **Gridlines** tab. Check the boxes labeled *Major Gridlines* for both the X and Y axes. Major gridlines for Y axis are checked by default.

Step 7: Choose the **Legend** tab and turn off the legend. (When the chart shows only a single curve, the legend doesn't tell you very much).

Step 8: Check the box labeled *As new sheet*, as this should be the last step of the Chart Wizard to indicate where the new chart should be placed.

In each of the following steps, you will be opening one of the Format dialog boxes and completing the formating procedure:

Step 1: Set the Y axis limits to −10,000 and 10,000. Select the Y axis object and choose the **Scale** tab to make this change.

Step 2: Move the Y axis to the left edge of the chart. Do this by selecting the X axis object and choosing the **Scale** tab. In the box labeled *Value (Y) axis crosses at*, type **−15.**

Step 3: Display the curve with a heavier line. Select the data series by clicking on the curved line. Choose the **Patterns** tab. Select a heavier weight line.

Step 4: Display the gridlines in gray rather than black. Double click on an X axis gridline and set the line color to light gray. Repeat for the Y gridlines.

After making these formatting changes, your chart should resemble Figure 4.27.

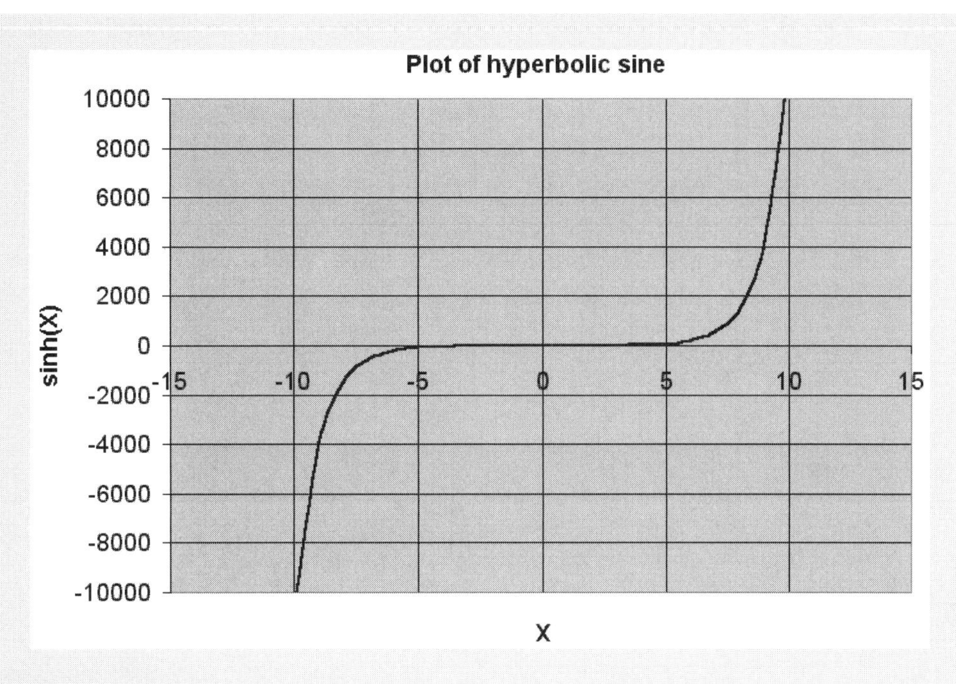

Figure 4.27. The hyperbolic sine function.

	A	B
1	X	sinh(X)
2	-10	-11013
3	-9	-4052
4	-8	-1490
5	-7	-548
6	-6	-202
7	-5	-74
8	-4	-27
9	-3	-10
10	-2	-4
11	-1	-1
12	0	0
13	1	1
14	2	4
15	3	10
16	4	27
17	5	74
18	6	202
19	7	548
20	8	1490
21	9	4052
22	10	11013

Figure 4.28. Evaluation of the hyperbolic sine function.

KEY TERMS

1, 2, 5 rule	chart objects	data series
error bars	gridlines	legend
major units	minor units	ticks
X axis	Y axis	

SUMMARY

In this chapter, you were shown how to create charts from an Excel worksheet. Examples were given of line charts and XY scatter plot, as these types of charts are frequently used in scientific and engineering applications. You were shown many of the wide variety of formatting options. Finally, several topics applicable to engineering charts—scaling axes and creating error bars—were presented.

Problems

1. Generate data points for the function

$$y = 4\sin(x) - x^2$$

for $x = -5.0, -4.5, \ldots, 4.5, 5.0$. Chart the results by using an XY scatter plot. Add appropriate title and axis labels.

2. The equation for the plot of a circle is

$$x^2 + y^2 = r^2,$$

where r is the radius of the circle. For a radius of 5, the Excel equation for y is

$$y = \text{SQRT}(5 - x\text{^}2).$$

Generate points for y for $x = 0.0, \ldots, 2.2$ in increments of 0.1 and plot the results. Why does your plot show only $\frac{1}{4}$ of a circle? Can you create more data points to plot a full circle?

3. The resolution of the data can dramatically change the appearance of a graph. If there are too few data points, the plot will not be smooth. If there are too many data points, the time and storage requirements become a burden. Generate two sets of data points for the following function:

$$f(x) = \sin(x).$$

For the first set use $x = 0.1, \ldots, 25.0$ in increments of 0.1 (250 data points). (You will definitely want to use the Fill Handle for this!) Generate another set with $x = 0, \ldots, 25$ in increments of 1 (25 data points). Plot both versions of $\sin(x)$. Does the plot with 25 data points give you a correct impression of the shape of the sine function?

4. A graph that uses logarithmic scales on both axes is called a log–log graph. A log–log graph is useful for plotting power equations, since they appear as straight lines. A power equation has the following form:

$$y = ax^{b}.$$

Table 4.2 presents data that are collected from an experiment that measured the resistance of a conductor for a number of sizes. The size (cross-sectional area) was measured in millimeters squared, and the resistance was measured in milliohms per meter. Create a scatter plot of these data.

TABLE 4.2 Resistance vs. Area of a Conductor

Area (mm^2)	Resistance (milliohms/meter)
0.009	2000.0
0.021	1010.0
0.063	364.0
0.202	110.0
0.523	44.0
1.008	20.0
3.310	8.0
7.290	3.5
20.520	1.2

5. Modify the X and Y axes of the chart created in the previous problem to use a logarithmic scale. From viewing the resulting scatter plot, what can you infer about the relationship between resistance and size of a conductor in this experiment?

6. Table 4.3 shows the average daily traffic flow at four different intersections for a five-year period. Create a worksheet and enter the table. Plot the data for intersections 1, 3, and 4, but not for intersection 2. Use different line types and markers for each of the three lines. Your graph should look like Figure 4.29.

TABLE 4.3 Average Daily Traffic Flow at Four Downtown Intersections

	Average Daily Traffic Flow (× 1,000)			
	Intersection #			
Year	**1**	**2**	**3**	**4**
1996	25.3	12.2	34.8	45.3
1997	26.3	14.5	36.9	48.7
1998	28.6	14.9	42.6	43.2
1999	29.0	16.8	50.6	46.9
2000	32.4	17.6	70.8	54.9
2001	34.8	17.9	82.3	60.9

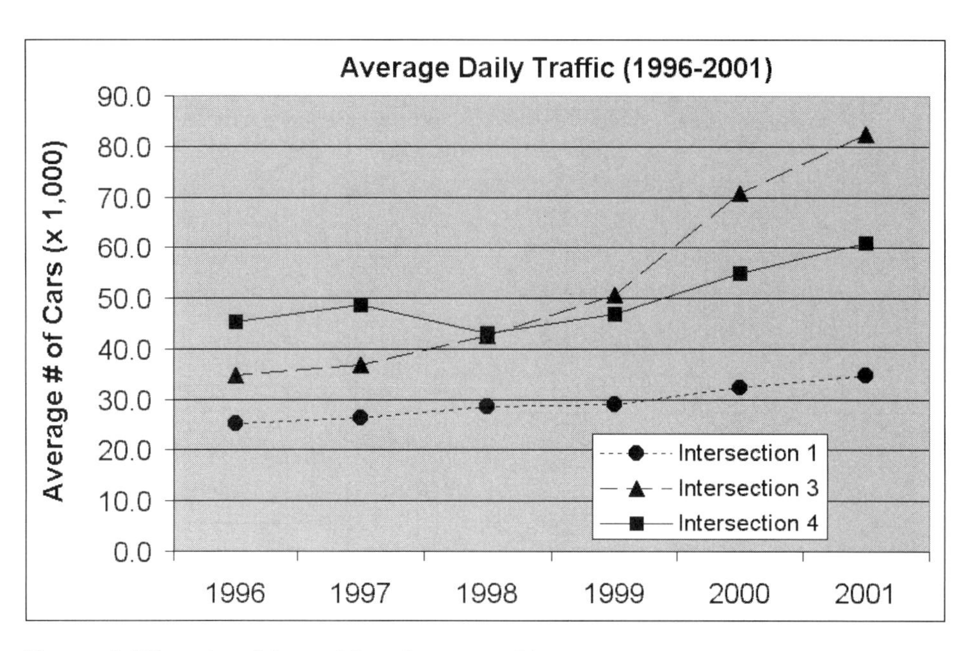

Figure 4.29 A plot of three of the columns in Table 4.3.

5

Performing Data Analysis

5.1 USING THE ANALYSIS TOOLPAK

An add-in package is available for Excel that includes a number of statistical and engineering tools. This package, called the *Analysis ToolPak*, can be used to shorten the time that it usually takes to perform a complex analysis.

The Analysis ToolPak is installed as a separate add-in to Excel. To determine whether the ToolPak is installed on your system follow these steps:

Step 1: Choose **Tools** from the Menu bar. If the Data Analysis command does not appear on the Tools menu, then the Analysis ToolPak has not been installed or has not been configured correctly.

Step 2: Load the add-in into Excel by choosing **Tools → Add-Ins** from the Menu bar. Check the box labeled *Analysis ToolPak* and click **OK**. If Analysis ToolPak does not appear on the Add-In menu, you must install it by running the Setup program on your Microsoft Office or Excel CD. After you have installed the ToolPak with the Setup program, return to this step.

Once the ToolPak is successfully installed, choose **Tools → Data Analysis** from the Menu bar. The Data Analysis dialog box will appear, as shown in Figure 5.1.

SECTION

OBJECTIVES

After reading this chapter, you should be able to:

- Access and use the Analysis ToolPak.
- Create a histogram.
- Calculate descriptive statistics for a data series.
- Calculate the correlation between two data series.
- Perform a linear regression analysis on a set of data.
- Calculate linear and exponential trends for data series.
- Project trendlines onto charts.
- Undertake the iterative solution of equations using the Goal Seeker.
- Perform constrained linear and nonlinear optimization using the Solver tool.

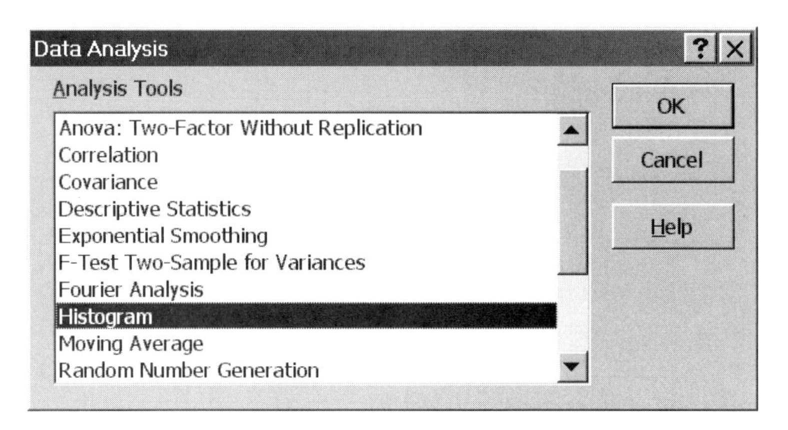

Figure 5.1. The Data Analysis dialog box.

Many analysis tools are included in the ToolPak. Each tool requires a set of input parameters in a specific format. These usually consist of an input range, an output range, and varying options. The results of the analysis are displayed in an output table. Additionally, some tools will generate a chart.

We will demonstrate how to use the Analysis ToolPak by using four of the available tools to

- Create a histogram
- Provide descriptive statistics
- Compute a correlation
- Perform a regression analysis

It is beyond the scope of this text to interpret the results of these statistical analyses or to explain the meaning of statistical terms such as confidence interval, residuals, R square, standard error, etc.

5.2 CREATING A HISTOGRAM

We will first demonstrate the use of the Analysis ToolPak with the creation of a histogram. A *histogram* is a graph of the frequency distribution of a set of data. The data are aggregated into classes and the classes are graphed in a bar chart. The width of each bar represents the range of a class, and the height of each bar represents the frequency of data within a particular range. A class range is sometimes called a *bin*, since the histogram effectively places each data point into a bucket or bin.

An example of the use of a histogram is to visualize the distribution of the tolerance errors of a test set of machine parts coming off of an assembly line. A glance at the histogram can tell the quality assurance (QA) department whether the tolerance errors are widely or narrowly disbursed. In addition, you can quickly see whether the tolerance errors are normally distributed or skewed.

The data in Figure 5.2 are the results of a test batch from an assembly line. The QA team collected 25 sample parts and measured their tolerances. The team noted that the test data typically range from about −5 to +5 thousandths of an inch from the correct size. The team created nine bins for the data to fall into.

To create a histogram of the data in Figure 5.2, using the Analysis ToolPak, perform the following steps:

	A	B	C	D
1	**Machine Parts Assembly Line 14**			
2			Test Batch #:	3201
3	**Tolerance**	**Bins**	Date:	02/01/02
4	0.34	-3.5	Units:	inches x 10
5	1.03	-2.5		
6	-1.26	-1.5		
7	3.13	-0.5		
8	-0.10	0.5		
9	0.02	1.5		
10	-0.01	2.5		
11	2.12	3.5		
12	-1.40	4.5		
13	1.24			
14	2.29			
15	-0.71			
16	-1.38			
17	-1.13			
18	-1.34			
19	0.03			
20	-0.03			
21	-0.56			
22	-0.04			
23	-2.55			
24	-0.01			
25	0.56			
26	0.78			
27	0.99			
28	1.12			

Figure 5.2. Tolerances of test batch of machine parts.

Step 1: Choose **Tools → Data Analysis** from the Menu bar. The Data Analysis dialog box will appear, as shown in Figure 5.1.

Step 2: Select **Histogram** from the menu. The Histogram dialog box will appear, as shown in Figure 5.3.

Step 3: Select the input range containing the test measurements (A3:A28).

Step 4: Select the bin range containing the bin boundaries (B3:B12). The bin ranges are defined when you list the boundary values in ascending order. A data point is determined to be in a particular bin if the value of the data point is less than, or equal to, the bin number and greater than the previous bin number. You can choose to omit the bin range, in which case, a set of evenly distributed bins between the data's minimum and maximum values is automatically created.

Step 5: Note that the selected ranges include the column headings. Check the box titled **Labels**, and the range headings will be used to label the histogram.

Step 6: The output can be directed to a specified range, a new worksheet, or a new workbook. Check the box titled **New Worksheet Ply**.

Step 7: Check the box titled **Chart Output**. Choose **OK**.

Step 8: The resulting frequency-distribution table and histogram are displayed in Figure 5.4. You can use any of the chart formatting options that you learned in the previous chapter to format the histogram.

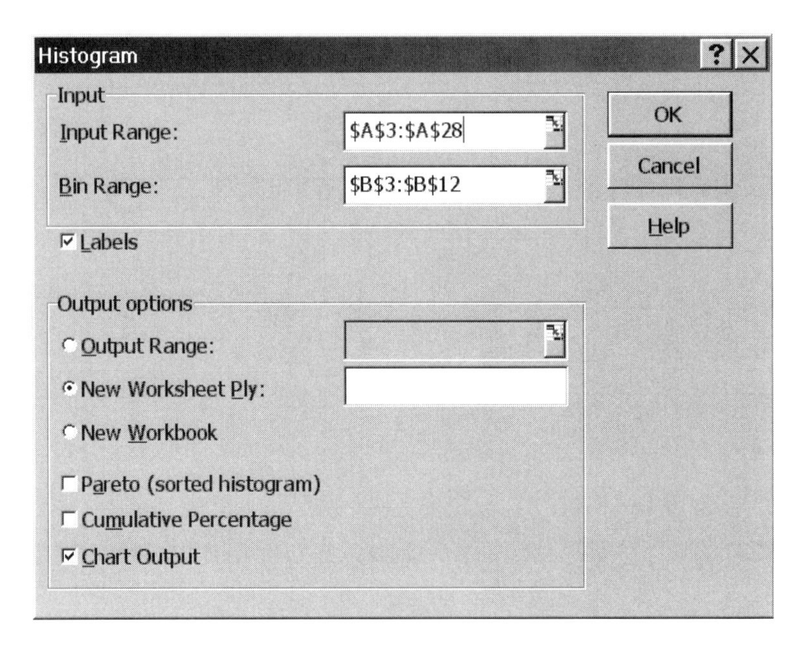

Figure 5.3. The Histogram dialog box.

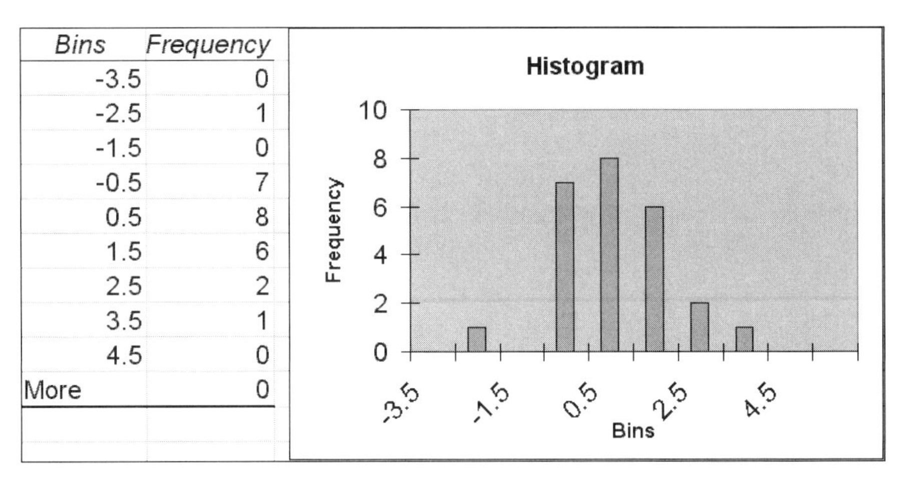

Bins	Frequency
-3.5	0
-2.5	1
-1.5	0
-0.5	7
0.5	8
1.5	6
2.5	2
3.5	1
4.5	0
More	0

Figure 5.4. Histogram of tolerances of tested machine parts.

Several other features are available on the Histogram dialog box. These include sorting the histogram (*Pareto*) and displaying a cumulative percentage (*Cumulative Percentage*).

The Help button on the Histogram dialog box displays detailed information about the input parameters and options for the Histogram tool. The Help button can be used to access unique help for each of the Analysis Tools.

5.3 PROVIDING DESCRIPTIVE STATISTICS

Excel provides a tool for computing common *descriptive statistics* (a form of describing the basic features of the data in a study) for sets of data. You could calculate each of

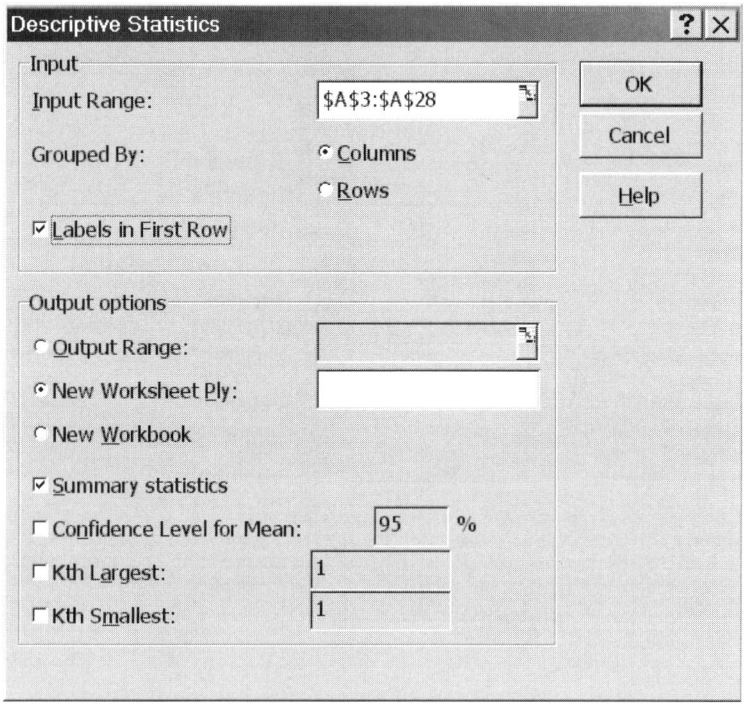

Figure 5.5. The Descriptive Statistics dialog box.

these values by using individual formulas from the formula menu. The data analysis tool conveniently aggregates the most common computations for you in a table.

As an example, we will go through the steps to compute the descriptive statistics of the machine-parts test data from Figure 5.2. To compute the descriptive statistics,

Step 1: Choose **Tools → Data Analysis** from the Menu bar. The Data Analysis dialog box will appear.

Step 2: Choose **Descriptive Statistics**. The Descriptive Statistics dialog box will appear, as shown in Figure 5.5.

Step 3: Choose the input range, including the column header (A3:A28).

Step 4: Check the box titled *Grouped by Columns*.

Step 5: For output options, check the boxes titled *New Worksheet Ply* and *Summary Statistics*.

Step 6: Choose **OK**. The results will appear in a new worksheet, as shown in Figure 5.6.

You are not expected to understand all of the terms in Figure 5.6. However, you probably recognize some of them. From these results, the QA team can see that the mean deviation from acceptable tolerance was close to zero. The data ranged from −2.55 to 3.13. An important result for the team is the variability of the test data. The standard deviation (1.304) is a measure of the data's variability. The standard deviation of the test set, along with the size of the test set, can be used to make predictions about how many of the items on the assembly line will eventually be rejected.

	A	B
1	*Tolerance*	
2		
3	Mean	0.125506494
4	Standard Error	0.260992976
5	Median	-0.01
6	Mode	-0.01
7	Standard Deviation	1.304964878
8	Sample Variance	1.702933332
9	Kurtosis	0.212116179
10	Skewness	0.295309542
11	Range	5.68
12	Minimum	-2.55
13	Maximum	3.13
14	Sum	3.137662338
15	Count	25

Figure 5.6. Descriptive statistics for the tested-machine parts.

5.4 COMPUTING A CORRELATION

The *correlation* of two data sets is a measure of how well the two ranges of data move together linearly. If large values of one set are associated with large values of the other, then a positive correlation exists. If small values of one set are associated with large values of the other (and vice versa), then a negative correlation exists. If the correlation is near zero, then the values in both sets are not linearly related.

Figure 5.7 shows the midterm grades and overall grade point averages (GPAs) for 20 students. We are interested in knowing whether there is a positive correlation between the midterm exam grades and the students' GPAs.

To calculate the correlation between the students' midterm grades and GPAs,

Step 1: Enter the data in Figure 5.7 into a worksheet.

Step 2: Choose **Tools → Data Analysis** from the Menu bar. The Data Analysis dialog box will appear.

Step 3: Choose **Correlation** from the Data Analysis dialog box. The Correlation dialog box will appear, as depicted in Figure 5.8.

Step 4: Select the input range to include the region (B1:C21) in Figure 5.7.

Step 5: Check the box titled *Grouped by Columns*.

Step 6: Check the box titled *Labels in First Row*.

Step 7: Check the box titled *New Worksheet Ply*. Then press **OK**.

The results are shown in Figure 5.9. Note that the correlation coefficient $r = 0.735437$, which indicates that there is a positive correlation between midterm grades and GPAs for this group of students.

	A	B	C
1	Student#	Mid Term	GPA
2	1	45	1.7
3	2	89	3.7
4	3	90	3.9
5	4	67	3.7
6	5	88	2.4
7	6	93	3.4
8	7	32	1.8
9	8	85	3.1
10	9	68	2.4
11	10	52	2.6
12	11	77	3.5
13	12	96	3.9
14	13	54	2.1
15	14	78	2.8
16	15	83	2.4
17	16	89	3.1
18	17	79	2.9
19	18	83	2.9
20	19	72	3.1
21	20	91	3.6

Figure 5.7. Midterm grades and GPAs for 20 students.

Correlation

Input

Input Range: $B$1:$C$21

Grouped By: ◉ Columns ○ Rows

☑ Labels in First Row

Output options

○ Output Range:

◉ New Worksheet Ply:

○ New Workbook

OK | Cancel | Help

Figure 5.8. The Correlation dialog box.

	A	B	C
1		*Mid Term*	*GPA*
2	Mid Term	1	
3	GPA	0.735437	1

Figure 5.9. The correlation between midterm grades and GPAs.

5.5 PERFORMING A LINEAR REGRESSION

The correlation is a measure of whether a linear relationship exists between two sets of data. However, the correlation coefficient does not tell us what the relationship is, merely that it exists. A *linear-regression analysis* is an attempt to find the relationship among variables and express the relationship as a linear equation.

The Analysis ToolPak performs linear-regression analysis by using the least squares method to fit a line through a set of observations. In this section, we will show you how to perform a regression that analyzes how a single dependent variable is affected by a single independent variable.

If the regression is a good fit to the data, then the regression allows us to make predictions of future performance. For example, we might make a prediction about a student's first-year college GPA based on the student's high school GPA. After having computed the relationship between a representative sample of previous students' high school GPAs and their first-year college GPAs, we would use the regression equation to predict how students would do in their first year in college.

If more than one independent variable is considered, then the analysis technique is called *multiple regression*. For example, the student's SAT score and IQ might both be tested as predictors, in addition to the high school GPA.

Since we know that there is a positive relationship between overall GPA and midterm exams in the data in Figure 5.7, let's compute a regression analysis on the same data. We would like to know whether we can predict a student's midterm grade given the student's GPA. We are making some strong assumptions about our collected data. These are assumptions about our sampling technique and the underlying distribution of the student GPAs. We will ignore these ramifications here, but you will study them if you take a statistics course.

To perform a regression analysis, perform the following steps:

Step 1: Choose **Tools → Data Analysis** from the Menu bar. The Data Analysis dialog box will appear.

Step 2: Choose **Regression** from the Data Analysis dialog box. The Regression dialog box will appear, as depicted in Figure 5.10.

Step 3: Select the input range to include both columns of data from Figure 5.7, grouped by columns. Include the header columns in your selection.

Step 4: Direct the output to a *New Worksheet Ply*.

Step 5: Check the output option labeled *Labels*, then press **OK**. The results are depicted in Figure 5.11.

Figure 5.11 presents a lot more information than you probably care to know. One of the important results is the significance of the F statistic (0.0002). This means that there is a very low probability that the results are from chance. The second most important numbers are the coefficients (Intercept and GPA). We use these to build a predictive equation:

$$\text{Predicted Midterm Grade} = 19.272 * \text{GPA} + 18.696.$$

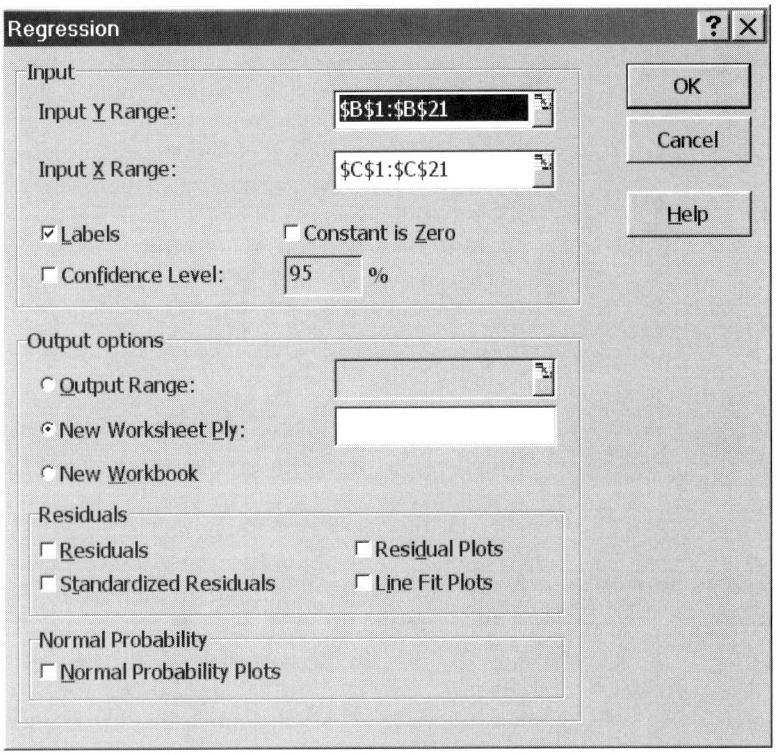

Regression

Input

Input Y Range: $B$1:$B$21

Input X Range: $C$1:$C$21

☑ Labels ☐ Constant is Zero

☐ Confidence Level: 95 %

OK

Cancel

Help

Output options

○ Output Range:

◉ New Worksheet Ply:

○ New Workbook

Residuals

☐ Residuals ☐ Residual Plots

☐ Standardized Residuals ☐ Line Fit Plots

Normal Probability

☐ Normal Probability Plots

Figure 5.10. The Regression dialog box.

	A	B	C	D	E	F
1	SUMMARY OUTPUT					
2						
3	*Regression Statistics*					
4	Multiple R	0.735				
5	R Square	0.541				
6	Adjusted R^2	0.515				
7	Std Error	12.266				
8	Observations	20				
9						
10	ANOVA					
11		*df*	*SS*	*MS*	*F*	*Significance F*
12	Regression	1	3190.55	3190.55	21.20	0.0002
13	Residual	18	2708.40	150.47		
14	Total	19	5898.95			
15						
16		*Coefficients*	*Std Error*	*t Stat*	*P-value*	
17	Intercept	18.696	12.648	1.478	0.157	
18	GPA	19.272	4.185	4.605	0.000	

Figure 5.11. A regression analysis of student GPAs.

5.6 TREND ANALYSIS

Trend analysis is the science of forecasting or predicting future elements of a data series based on historical data. Trend analysis is used in many areas, such as financial forecasting, epidemiology, capacity planning, and criminology. Excel has the ability to calculate linear and exponential growth trends for data series. Excel can also calculate and display various trendlines for charts.

A *linear trend* consists of fitting known x and y values to a linear equation. Unknown y values may then be predicted by extending the x values and using the equation to calculate new y values. This is called *linear extension*. An *exponential trend* can be created in a similar manner, except that the data are fit to an exponential equation. The graph of a trend is called a *trend line*.

5.6.1 Trend Analysis with Data Series

A trend analysis can either extend or replace a series of data elements. The simplest method for extending a data series with a linear regression is to drag the Fill Handle past the end of the data series. For example, Figure 5.12 shows the number of occurrences of a hypothetical disease for the years 1993 to 1996.

By assuming a linear rate of increase in the disease, the number of occurrences can be estimated for 1997, 1998, and 1999.

To calculate the *linear trend*, select the known data (A4:D4), then drag the Fill Handle to the right so that the fill box covers cells (E4:G4). When you release the mouse, cells (E4:G4) will contain the data elements predicted by a linear regression of the original data.

The Fill Series command can be used for somewhat more sophisticated trend analysis. To extend or replace a data series by using the Fill Series command, first select the region of data over which the analysis is to occur. This includes the original data and the new cells that are to hold the predicted data. Let's try some of the Fill Series functions, using the sample data in Figure 5.13:

Step 1: Create a worksheet with the titles and headers in Figure 5.13.

Step 2: Type the numbers 1.1, 1.9, 3.0, 3.8 into cells (B4:E4).

To extend the known values with a linear regression and leave the original values unchanged,

Step 1: Copy cells (B4:E4) to Row 6. Select the Row-6 data points, including the unknown points (B6:H6).

Step 2: Choose **Edit → Fill → Series** from the Menu bar. The Series dialog box will appear, as shown in Figure 5.14.

	A	B	C	D	E	F	G
1	Occurrence of Disease X (in thousands)						
2	Known				Projected		
3	1993	1994	1995	1996	(1997)	(1998)	(1999)
4	1.1	1.9	3.0	3.8	4.8	5.7	6.6

Figure 5.12. Disease data and projected disease occurrences.

	A	B	C	D	E	F	G	H
1		Occurrence of disease X (in thousands of cases)						
2								
3		1993	1994	1995	1996	1997	1998	1999
4	Original Known Values	1.1	1.9	3.0	3.8			
5								
6	Linear Extension	1.1	1.9	3.0	3.8	4.8	5.7	6.6
7								
8	Linear Replacement	1.1	2.0	2.9	3.8	4.8	5.7	6.6
9								
10	Exponential Replacement	1.2	1.8	2.7	4.1	6.3	9.5	14.5
11								
12	TREND() function	1.1	1.9	3.0	3.8	4.8	5.7	6.6

Figure 5.13. Examples of trend-analysis options.

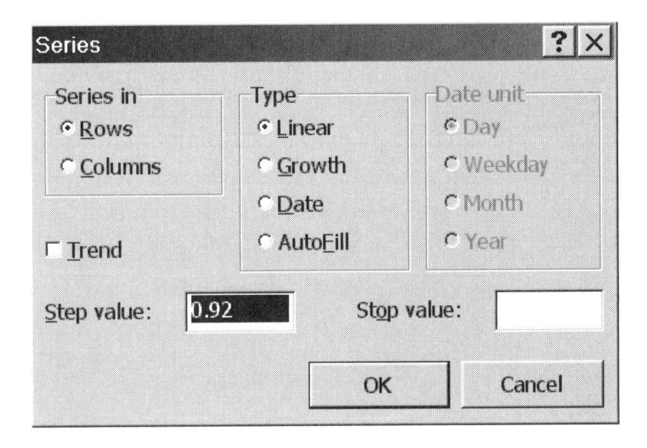

Figure 5.14. The Series dialog box.

Step 3: Note that a *step value* (the amount by which a series is increased or decreased) of one has been calculated by Excel. The step value can be modified manually to set the increment value for x in the linear equation

$$y = mx + b.$$

A *stop value* (the value at which the series is to end) may be entered if you want to set an upper limit to the trend.

Step 4: Check the box labeled *Series in rows*.

Step 5: Check the box labeled *AutoFill*.

Step 6: Uncheck all other boxes. The results are depicted in Row 6 in Figure 5.13 (labeled *Linear Extension*). Note that these results are identical to the results obtained by dragging the Fill Handle in Figure 5.12.

To calculate a linear trend and replace the original data values with best fit data,

Step 1: Copy cells (B4:E4) to Row 8. Select the Row-8 data points, including the unknown points (B8:H8).

Step 2: Choose **Edit → Fill → Series** from the Menu bar. The Series dialog box will appear.

Step 3: Check the box labeled *Series in rows*.

Step 4: Check the box labeled *Linear*.

Step 5: Check the labeled *Trend*. The trendline is no longer forced to pass through any of the original data points. Note that the data for 1994 and 1995 have been modified. The results are depicted in Row 8 in Figure 5.13 (labeled *Linear Replacement*).

To create a trend, using exponential growth series,

Step 1: Copy cells (B4:E4) to Row 10. Select the Row-10 data points, including the unknown points (B10:H10).

Step 2: Choose **Edit → Fill → Series** from the Menu bar. The Series dialog box will appear.

Step 3: Check the box labeled *Series in rows*.

Step 4: Check the box labeled *Growth*.

Step 5: Check the box labeled *Trend*. The original data values are replaced with a growth curve, and the trendline is not forced to pass through any of the original data points. The results are depicted in Row 10 in Figure 5.13 (labeled *Exponential Replacement*).

5.6.2 Trend-Analysis Functions

Excel provides two trend-analysis functions, one for linear trend calculation and another for calculating exponential trends. These are useful if the known dependent data may change and the trendline must be recalculated frequently. The linear trend function TREND uses the least squares method for its calculation. The syntax for TREND is

TREND (*Known_y's, Known_x's, New_x's, Const*).

The arguments are:

- **Known_y's.** The known y values are the known dependent values in the linear equation $y = mx + b$. In Figure 5.13, the known y values are 1.1, 1.9, 3.0, and 3.8 in the range (B4:E4).
- **Known_x's.** The known x values are the values of the independent variable for which the y values are known. In Figure 5.13, these are the values 1993, 1994, 1995, and 1996 in the range (B3:E3). If the known x values are omitted, then the argument is assumed to be $\{1, 2, 3, 4 \ldots\}$.
- **New_x's.** The new x values are the values of the independent variable for which you want new y values to be calculated. If you want the predictions for years 1997 to 1999, then select the range (F3:H3). If you want to calculate the linear trend for the whole time span (1993–1999), then select the range (B3:H3).
- **Const.** If the *const* argument is set to FALSE, then b is set to zero, so the equation describing the relationship between y and x becomes $y = mx$. If the const argument is set to TRUE or omitted, then b is computed.

To use the TREND function with our example,

Step 1: Copy the original 4 data points to cells (B12:E12).

Step 2: Select a region in which to place the results (F12:H12). The region should be the same size as the number of *New_x's* to be calculated.

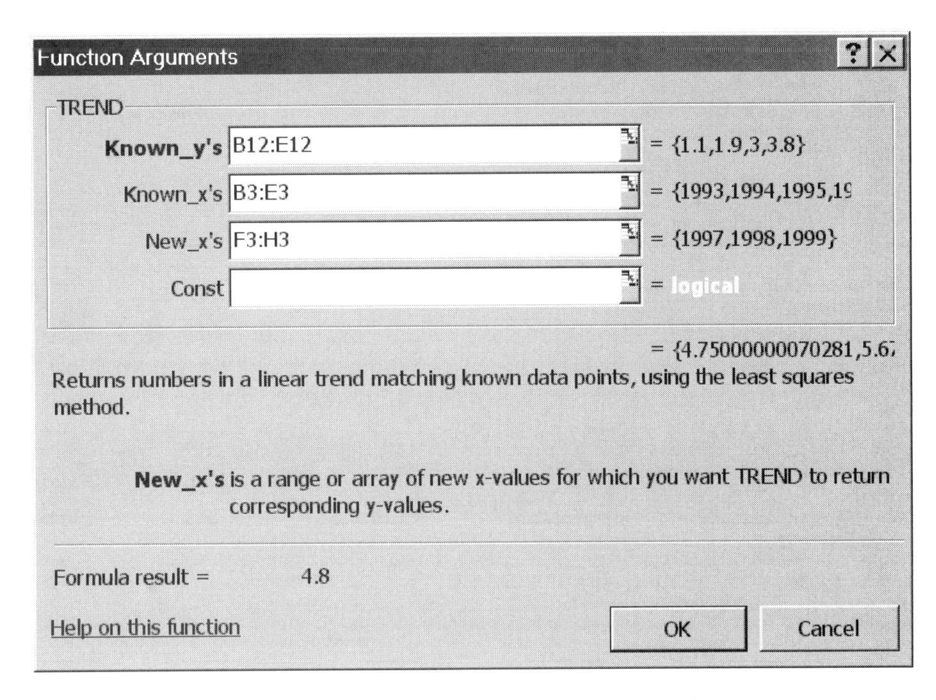

Figure 5.15. The Function Arguments dialog box for the **TREND** function.

Step 3: Choose the Insert Function icon and select the **Statistical** category. Choose the **TREND** function from the function list. The Function Arguments dialog box will appear, as shown in Figure 5.15.

Step 4: Select or type the range (B12:E12) for the *Known_y's*.

Step 5: Select or type the range (B3:E3) for the *Known_x's* (the years 1993–1996).

Step 6: Select or type the range (F3:H3) as the *New_x's* (the years 1997–1999).

Step 7: Don't click **OK**. Since we want the output to be an array, rather than a single value, press **Ctrl + Shift + Enter**.

Step 8: The results are depicted in Row 12 of Figure 5.13 (labeled *TREND() function*).

An alternate method for performing the same task is to

Step 1: Select the output cells (F12:H12).

Step 2: Type the following formula: =TREND(B12:E12,B3:E3,F3:H3)

Step 3: Press **Ctrl + Shift + Enter**.

There is a corresponding Excel function for producing an exponential trend, named GROWTH. The arguments for the GROWTH function are similar to the arguments for TREND.

5.6.3 Trend Analysis for Charts

Excel will calculate and display graphic representations of trends on a chart with the use of trendlines. Five types of regression lines can be added, or a moving average can be calculated. Each type of trendline is described in Table 5.1.

TABLE 5.1 Excel Trendline Types

Trendline Type	Formula
Linear	Calculates the least squares fit by using $y = mx + b$. (m is the slope and b is a constant.)
Logarithmic	Calculates least squares by using $y = c \cdot \ln(x) + b$. (c and b are constants.)
Polynomial	Calculates least squares for a line by using $y = b + c_1x + c_2x^2 + \ldots + c_nx^n$. ($b, c_1, c_2, \ldots, c_n$ are constants, the order can be set in the Add Trendline dialog box, and the maximum order is 6.)
Exponential	Calculates least squares by using $y = ce^{bx}$. (c and b are constants.)
Power	Calculates least squares by using $y = cx^b$. (c and b are constants.)
Moving Average	Calculates the series of moving averages by using $F_{(t+1)} = \frac{1}{N} \sum_1^N A_{t-j+1}$. Keep in mind that each data point in a moving average is the average of a specified number of previous data points. (N is the number of prior periods to average, A_t is value at time t, and F_t is the forecasted value at time t.)

Trendlines cannot be added to all types of charts. For example, trendlines cannot be added to data series in pie charts, 3-D charts, stacked charts, or doughnut charts. Trendlines can be added to bar charts, XY scatter plots, and line charts. If a trendline is added to a chart and the chart type is subsequently changed to one of the exempted types, then the trendline is lost.

To create a trendline, first generate a chart of an acceptable type. As an example, we will make an XY scatter chart by using the data in Figure 5.16. These data represent the growth rate of a certain type of algae in the Great Salt Lake. It has been predicted that, if left undisturbed, the algae would multiply exponentially.

	A	B
1	Algae Growth Rate	
2	Day	Population (millions)
3	1	1.00
4	2	1.12
5	3	1.92
6	4	2.65
7	5	4.12
8	6	6.41
9	7	8.66
10	8	14.36
11	9	23.34
12	10	34.22

Figure 5.16. Growth rate of algae.

Here are the steps you must use:

Step 1: Create a worksheet containing the data in Figure 5.16.

Step 2: Create an XY scatter chart from the data in Figure 5.16 by using the Chart Wizard.

Step 3: Create a proper title and labels for the chart, as shown in Figure 5.17.

Step 4: Select the chart with the mouse.

Step 5: Choose **Chart → Add Trendline** from the Menu bar. The Add Trendline dialog box will appear.

Step 6: Choose the **Type** tab, as shown in Figure 5.18.

Step 7: Choose **Exponential** from the box labeled *Trend/Regression type*. This will enable us to see how well an exponential-growth line fits over the collected data.

Step 8: Choose the name that you gave your series from the list labeled *Based on series*. (Our example uses **Algae Growth Rate**.)

Step 9: Now choose the **Options** tab on the Add Trendline dialog box, as depicted in Figure 5.19. (The first box under the **Options** tab gives you the option of naming the trendline.)

Step 10: Check the first box under the **Options** tab titled *Automatic*. Note that this box gives the option of naming of the trendline. However, we choose to let Excel name the trendline. By checking the *custom* box, we would have to write in our choice for the trendline's name.

Step 11: Leave the second box set to zero. This box can be set to forecast data points prior to, or after, the input data.

Step 12: Leave the box labeled *Set intercept* checked. This box allows you to set the y intercept to a constant value. However, if it is not checked, then the intercept will be calculated by Excel.

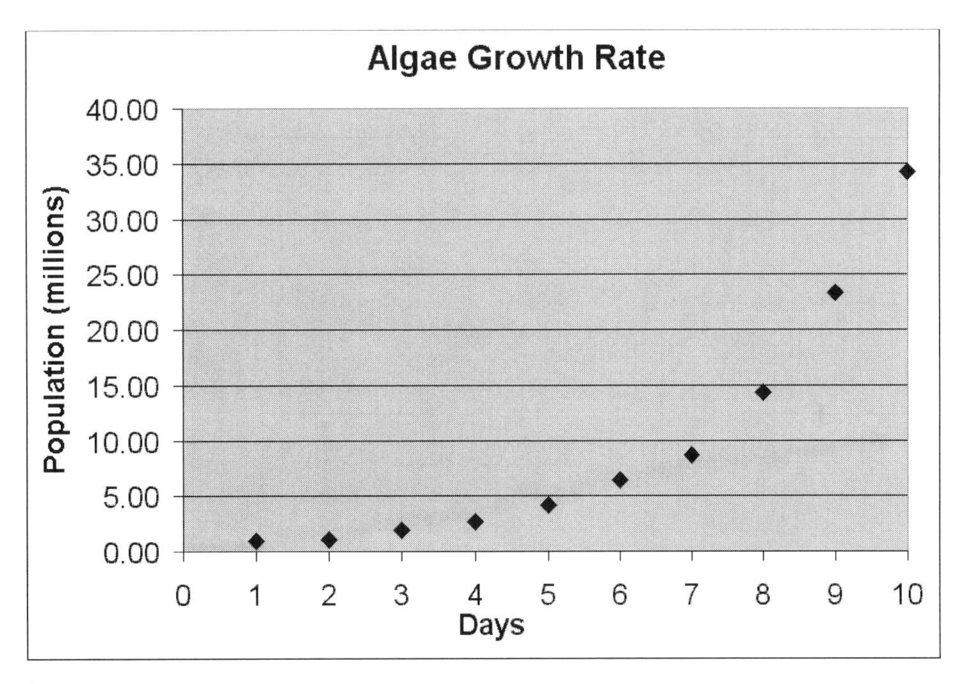

Figure 5.17. A chart of the algae growth-rate data.

Add Trendline
Type
Options
Trend/Regression type
Linear
Logarithmic
Polynomial
Order:
2
Power
Exponential
Moving Average
Period:
2
Based on series:
Algae Growth Rate
OK
Cancel

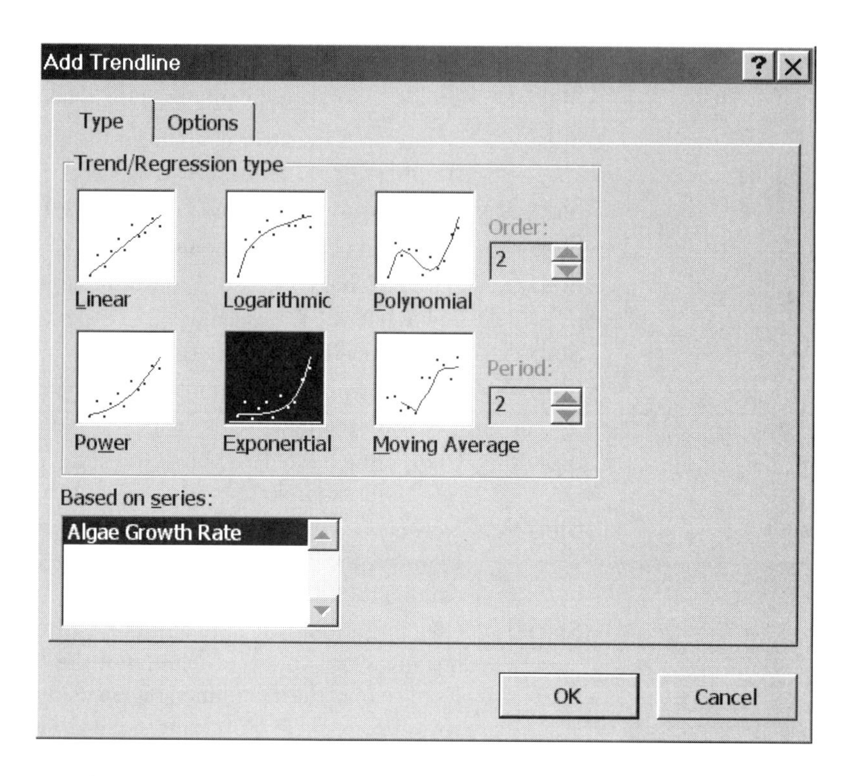

Figure 5.18. The Type tab of the Add Trendline dialog box.

Add Trendline
Type
Options
Trendline name
Automatic:
Expon. (Algae Growth Rate)
Custom:
Forecast
Forward:
0
Units
Backward:
0
Units
Set intercept =
0
Display equation on chart
Display R-squared value on chart
OK
Cancel

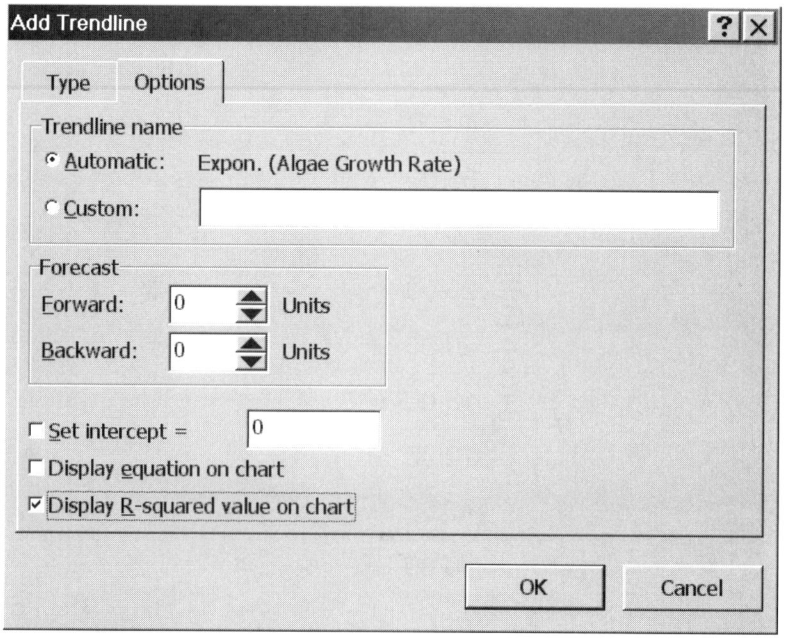

Figure 5.19. The Options tab of the Add Trendline dialog box.

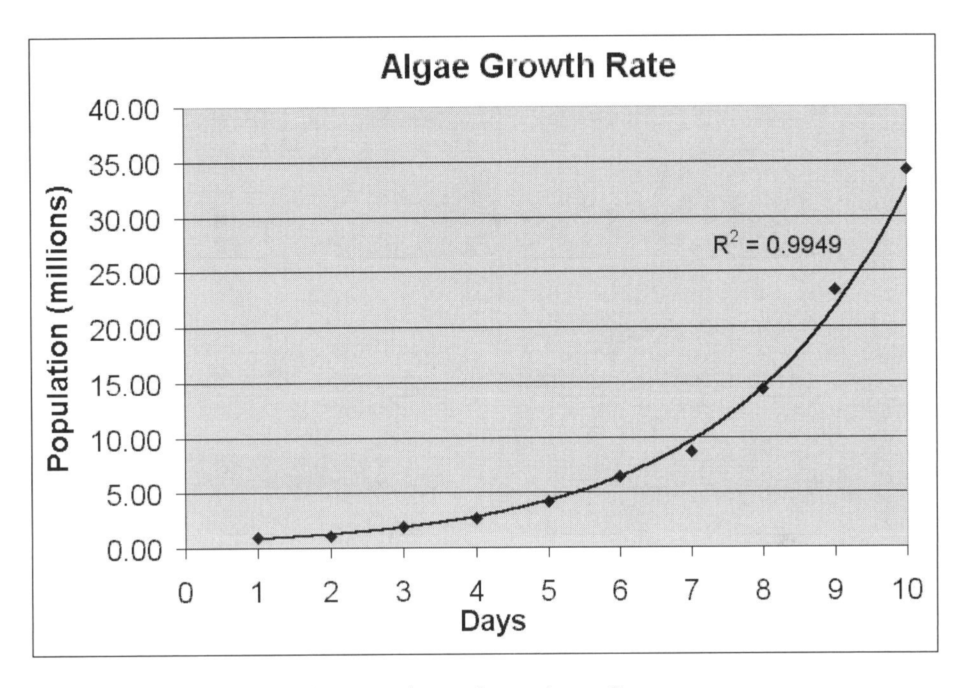

Figure 5.20. Chart of the algae growth-rate data with trendline.

Step 13: Check the box labeled *Display R-squared value on chart*, then click **OK**. The R^2 value is a measure of how well the data fit the trendline. (1.0 is a perfect fit.)

Step 14: The resulting chart and trendline should resemble Figure 5.20.

5.7 USING THE GOAL SEEKER

The *Goal Seeker* is used to find the input values of an equation when the results are known. It takes an initial guess for an input value and uses iterative refinement to attempt to locate the real input value.

An example is to use the Goal Seeker to find the solution of a polynomial equation:

$$f(x) = 3x^3 + 2x^2 + 4 = 0.$$

The equation $f(x)$ has a real solution, which is approximately -1.3. We can use the initial guess of 1 as a seed for the Goal Seeker, which will then attempt to converge on a more accurate value for x. The Goal Seeker does not always converge. With some functions, the initial guess must closely approximate the real solution.

To see how the Goal Seeker works, perform the following steps:

Step 1: Create a worksheet that resembles Figure 5.21. Note the coding for $f(x)$ in the Formula box. The formula should be placed in cell C3.

Step 2: Place 1, which is our initial guess for x, in cell C4. This results in $f(x) = 9$.

Step 3: Choose **Tools → Goal Seek** from the Menu bar. The Goal Seek dialog box will appear, as depicted in Figure 5.22. We want the formula to result in zero, using one (1) as an initial guess. The box titled *Set cell* should contain the formula to solve B3.

Step 4: Place a zero (0) in the box titled *To value*, since we want to know the value of x when the formula equals zero.

B3 | f_x =3*B4^3+2*B4^2+4

	A	B	C	D
1	Solution for f(x) = $3x^3 + 2x^2 + 4$			
2				
3	Formula =	9		
4	x =	1		
5				

Figure 5.21. Solving a polynomial equation, using the Goal Seeker.

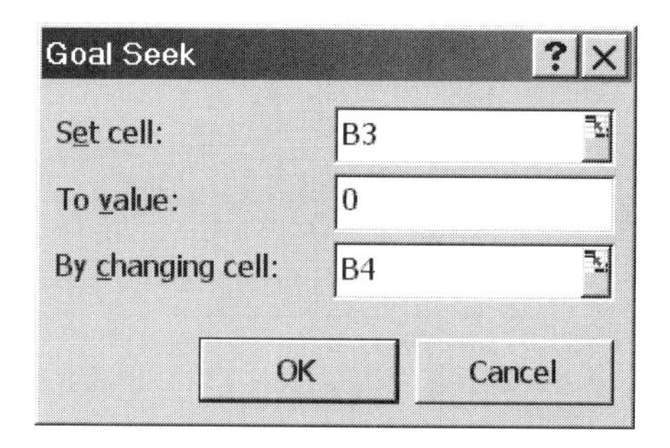

Figure 5.22. The Goal Seek dialog box.

Step 5: Insert B4 in the box titled *By changing cell* as the box should hold the cell containing our initial guess of the solution.

Step 6: Choose **OK**. The resulting value, $x = -1.373468173$, produces a solution for $f(x) \approx 0.0000375$, which is pretty close to zero.

5.8 USING SOLVER FOR OPTIMIZATION PROBLEMS

The solution of constrained nonlinear optimization problems is an important task for engineers in the petroleum, chemical, defense, financial, agriculture, and process-control industries. The Excel Solver may be the most widely used optimization software in the world.

What is an optimization problem? Many engineering problems have more than one solution. Engineers choose among a range of possible solutions by applying limits to the input parameters of the problem.

Given the limitations, the problem becomes one of finding a minimum (or maximum) solution. For example, you may want to minimize the cost of the production of widgets given the limitations of staff hours, availability of raw materials, power consumption, etc.

The equation that is to be maximized (or minimized) is called the *objective function*. The limitations to the input parameters of the objective function are called *constraints*. Problems of this type—finding a minimum or maximum given multiple constraints—are called *optimization* problems.

5.8.1 Introduction to Microsoft Excel Solver

The nonlinear optimization method used by Excel Solver is called the Generalized Reduced Gradient (GRG2) method. The GRG2 code was developed by Leon Lasdon, University of Texas at Austin, and Allan Waren, Cleveland State University. John Watson and Dan Fylstra, of Frontline Systems, Inc., implemented the methods used for linear and integer problems.

Solver must be installed as an Excel Add-In. To see if Solver is installed on your system,

Step 1: Choose **Tools** from the Menu bar and look for the **Solver** menu item.

Step 2: Choose **Tools** → **Add-Ins** from the Menu bar if Solver is not present as a menu item. Choose **Solver** from the Add-Ins dialog box, and click **OK**.

If Solver does not appear on the Add-Ins list, then it must be installed from the Microsoft Excel or Microsoft Office installation CD.

5.8.2 Setting Up an Optimization Problem in Excel

The most difficult part of solving an optimization problem is setting up the objective function and identifying the constraints. An objective function takes the form

$$y = f(x_1, x_2, \ldots, x_n).$$

The independent variables (x) are limited by m constraints, which take the form

$$c_i(x_1, x_2, \ldots, x_n) = 0 \text{ for i} = 1, 2, \ldots, m.$$

The constraints may also be expressed as inequalities. Excel Solver can handle both linear and nonlinear constraints. However, nonlinear constraints must be continuous functions. Although the constraints are expressed as functions, they are always evaluated within a range of precision called the *tolerance*. A constraint such as $x_1 < 0$, if the tolerance is large enough, may be evaluated as TRUE when $x_1 = 0.0000003$.

Nonlinear optimization problems may have multiple minima or maxima. In a minimization problem, all solutions, except the absolute minimum, are called *local minima*. The solution that is chosen by Solver is dependent on the initial starting point for the solution. The initial guess should be as close to the real solution as possible. These are problems with optimization in general, not just with Excel Solver.

This has probably been rather confusing. A couple of examples will make things clearer. Let's proceed by setting up examples of both a linear and a nonlinear optimization problem.

5.8.3 Linear Optimization Example

Assume that you wish to maximize the profit for producing widgets. The widgets come in two models: economy and deluxe. The economy model sells for $49.00, and the deluxe model sells for $79.00. The cost of production is determined primarily by labor costs, which are $14.00 per hour. The union limits the workers to a total of 2,000 hours per month. The economy widget can be built in 3 person-hours and the deluxe widget can be built in 4 person-hours. The management believes that it can sell up to 600 deluxe widgets per month and up to 1,200 economy widgets. Since you have a limited workforce, the main variable under your control is the ability to balance the number of economy units vs. deluxe units that are built. Your job is to determine how many economy widgets and how many deluxe widgets should be built to maximize the company's profit.

The independent variables are as follows:

w_1 = the number of economy widgets produced each month;
w_2 = the number of deluxe widgets produced each month.

The target that you wish to maximize is the profit, p. It is described mathematically by the objective function

$$p = (\$49 - (3 \text{ person-hours} * \$14/\text{hour}))\, w_1 + (\$79 - (4 \text{ person-hours} * \$14/\text{hour}))\, w_2 = 7\, w_1 + 23\, w_2$$

The constraints can be expressed mathematically as limitations on w_1 and w_2. The maximum number of widgets to be produced is limited by both sales and labor availability. The sales limitations (imposed by management) can be expressed as

$$w_1 \leq 1{,}200 \text{ widgets}$$

and

$$w_2 \leq 600 \text{ widgets.}$$

The availability of labor is limited by the labor union and can be expressed as

$$3\, w_1 + 4\, w_2 \leq 2000.$$

Finally, the general constraint of nonnegativity is imposed on w_1 and w_2, since you cannot produce a negative number of widgets:

$$w_1 + w_2 \geq 0$$

Figure 5.23 shows how to set up the widget problem in a worksheet. Cells D5 and D6 have been named *w_1* and *w_2*, respectively, to make the formulas more readable. Note the underlines in the names. The names *w1* and *w2* cannot be used, since Excel reserves these for cell identifiers. The formulas are displayed in the cells for readability. To display formulas, instead of their results, choose **Tools → Options → View** from the Menu bar. Check the box labeled *Formulas*.

	A	B	C	D
1	Widget Profit Optimization Worksheet			
2				
3				
4		Independent variables		
5			Ecomony widgets/month w1=	0
6			Deluxe widgets/month w2=	0
7				
8		Objective function		
9			Profit(p)	=7*w_1 + 23*w_2
10				
11		Constraints		
12			Labor constraint	=3*w_1 + 4*w_2

Figure 5.23. Worksheet for linear-optimization example.

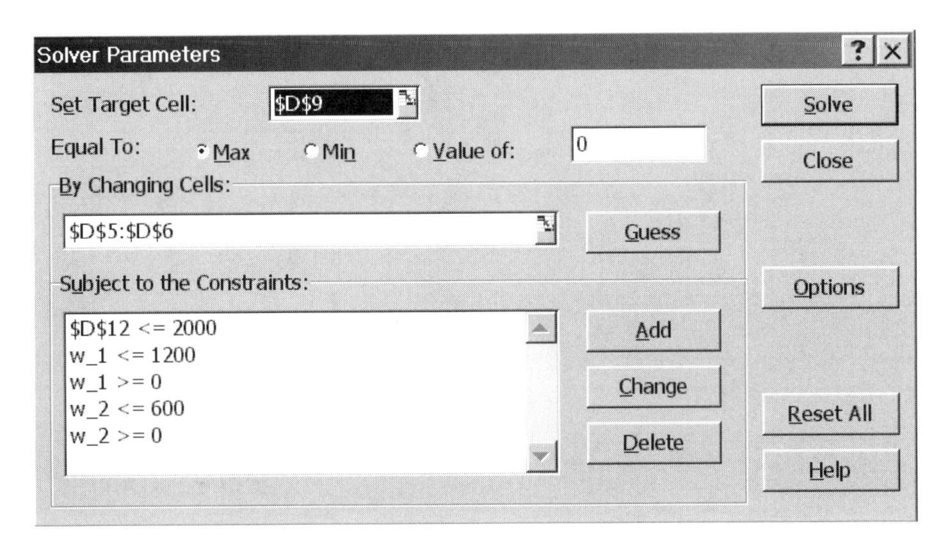

Figure 5.24. The Solver Parameters dialog box.

You have already completed the hardest part of using the Solver, which is setting up the worksheet. To run the Solver,

Step 1: Choose **Tools** → **Solver** from the Menu bar. The Solver Parameters dialog box will appear, as depicted in Figure 5.24. The box labeled *Set Target Cell* should contain the cell holding the objective function (the profit function). In our example, that is cell ($D$9).

Step 2: Check the box labeled *Max*, since you want to maximize profit. The box labeled *By Changing Cells* should contain the input parameters *w_1* and *w_2*. These are cells ($D$5) and ($D$6) in the example.

Step 3: Add each of the following five constraints:

    $D$12 <= 2000;
    w_1 <= 1200;
    w_1 >= 0;
    w_2 <= 600;
    w_2 >= 0.

Step 4: Choose **Options**, when you are satisfied with the constraint selection.

Step 5: The Solver Options dialog box will appear, as depicted in Figure 5.25. Accept most of the default options in the Solver Options dialog box. These should not be changed unless you understand the GRG2 and simplex methods used to implement Solver.

Step 6: Make sure that the box labeled *Assume Linear Model* is checked, since this is a linear problem.

Step 7: Choose **OK** to return to the Solver Parameters dialog box.

Step 8: Choose **Solve** to begin the computation. The Solver Results dialog box will appear, as depicted in Figure 5.26. It should state that Solver has found a solution. There are three types of reports available from Solver: **Answer**, **Sensitivity**, and **Limits**.

Step 9: Select all three types of reports.

Step 10: Check the box labeled *Keep Solver Solution* and then press **OK**.

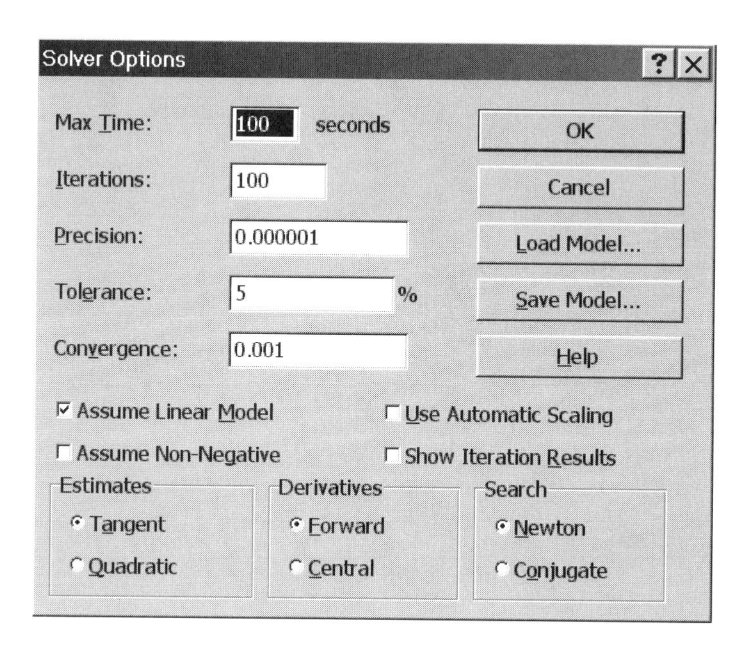

Figure 5.25. The Solver Options dialog box.

Figure 5.26. The Solver Results dialog box.

The worksheet that you created will now contain modified values for the input parameters and objective function. In addition, three new worksheets will be created for the Answer, Sensitivity, and Limits reports.

The Answer report summarizes the initial and final values of the input parameters and the optimized variable. The Sensitivity report describes information about the marginal effects of making small changes in the constraints. Sometimes, a small constraint change can make a large difference in the output. For nonlinear models, these are called *Lagrange multipliers*. For linear models, these are called either *dual values* or *shadow prices*. The Limits report shows the effect on the solution as each input parameter is set to its minimum or maximum limit.

The resulting spreadsheet from our example is shown in Figure 5.27. A brief look at this worksheet shows that the maximum profit, $11,500 per month, is achieved by producing only deluxe widgets. Only 500 deluxe widgets can be produced per month, but the company can sell 600 per month; thus, a limiting constraint is the available labor pool.

	A	B	C	D
1	Widget Profit Optimization Worksheet			
2				
3				
4	Independent variables			
5	Ecomony widgets/month w1=			0
6	Deluxe widgets/month w2=			500
7				
8	Objective function			
9			Profit(p)	$11,500
10				
11	Constraints			
12			Labor constraint	2000

Figure 5.27. Results of linear optimization.

As manager, you can easily modify the constraints and rerun Solver to see the effect. You could rapidly view the effect on profit of hiring more laborers, modifying prices, or adjusting the widget mix.

5.8.4 Nonlinear Optimization Example

As an example of nonlinear optimization, we will use an optimization problem for which the solution is obvious. This will familiarize you with the process of setting up a nonlinear optimization problem and convince you that the results are correct.

The objective function that we wish to minimize is

$$y = 100(x_2 - x_1^2)^2 + (1 - x_1)^2,$$

with the nonnegativity constraints

$$x_1 \geq 0$$

and

$$x_2 \geq 0$$

Since the terms $(x_2 - x_1^2)^2$ and $(1 - x_1)^2$ must be positive for real numbers x_1 and x_2, we know the answer. The minimum y is zero when $x_1 = 1$ and $x_2 = 1$. The worksheet for this example is shown in Figure 5.28.

Complete the following steps to use the Solver to perform nonlinear optimization:

Step 1: Choose **Tools → Solver** from the Menu bar. The Solver Parameters dialog box will appear.

Step 2: The box labeled *Set Target Cell* should contain the cell holding the objective function (the formula). In our example, that is cell ($C$8).

Step 3: Check the box labeled *Min* since you want to find the minimum of the function.

Step 4: The box labeled *By Changing Cells* should contain the input parameters *x*_1 and *x*_2. These are cells ($C$4) and ($C$5) in the example.

	A	B	C	D
1		Example of non-linear optimization		
2				
3		Input Parameters		
4		x1=	0	
5		x2=	0	
6				
7		Objective Function		
8		y=	=100*(x_2 - x_1^2)^2 + (1-x_1)^2	

Figure 5.28. Worksheet for nonlinear-optimization example.

Step 5: If you named cells (C4) and (C5) as *x*_1 and *x*_2, respectively, add the two following constraints:

x_1 >= 0;
x_2 >= 0;

otherwise, use the cell references in the constraints.

Step 6: Select **Options** and make sure that the box labeled *Assume Linear Model* is not checked.

Step 7: Click **OK** to exit the Options dialog box. Click **Solve** to start the Solver.

The results produced by Solver are good approximations of the true minimum:

$$x_1 = 0.999977;$$
$$x_2 = 0.999962;$$
$$y = 0.000000000649784.$$

There are other local minima for this objective function. Try setting the initial parameters to $x_1 = 3$ and $x_2 = 5$ and rerunning the Solver. The results produced by Solver show that the algorithm is stuck in a local minimum:

$$x_1 = 1.639202;$$
$$x_2 = 2.668455;$$
$$y = 0.44290084.$$

APPLICATION!—YIELD STRENGTH OF MATERIALS

Materials science is an important field of study for engineers, which covers the electronic, optical, mechanical, chemical, and magnetic properties of metals, polymers, composite materials, and ceramics.

Crystalline materials are subject to *slip deformation* when a shear stress is applied to the material. The deformation occurs when atomic planes slide along the directions of densest atomic packing, as shown in Figure 5.29.

When crystalline materials, such as metals and alloys, are formed by cooling molten metal, separate crystals form in the melt and grow together. The boundaries between the growing crystals form barriers to slip deformation, increasing the observed *yield strength* of the metal. The relationship between grain size (i.e., the size of the individual crystals in the metal) and observed yield strength, σ_y, is described by the Hall–Petch equation,

$$\sigma_y = \sigma_o + k_y \frac{1}{\sqrt{d}},$$

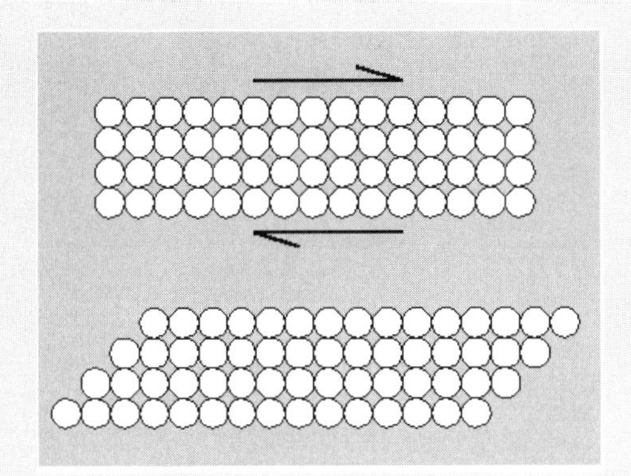

Figure 5.29. Slip deformation.

where σ_o is the yield strength of the pure metal (i.e., single crystal, $d = \infty$). The value of the proportionality factor, k_y, depends on the material and can be obtained by regression analysis from experimental data. (See Figure 5.30.)

For the carbon-steel data shown here, the coefficients for the Hall–Petch equation can be found by using Excel:

$$\sigma_y = 60.466 \frac{MN}{m^2} + 689.49 \frac{MN\,\mu m^{0.5}}{m^2} \frac{1}{\sqrt{d}}.$$

The k_y value would typically be reported as $k_y = 0.689 \frac{MN}{m^{3/2}}$.

HOW'D THEY DO THAT?

The Hall–Petch equation is an example of an equation that can be written in linear form

$$y = ax + b,$$

where a is the slope of the line through the data values and b is the y-intercept. Comparing terms with the Hall–Petch equation, you see that

$$\sigma_y = y;$$
$$\sigma_o = b;$$
$$\frac{1}{\sqrt{d}} = x.$$

A plot of $\frac{1}{\sqrt{d}}$ on the x axis against σ_y on the y axis shows the desired linear relationship. (See Figure 5.31.)

We can use Excel to plot a trendline and calculate the equation of the line through the data, using the following steps:

Step 1: Create a worksheet by using the table in Figure 5.30. Add a third column to your table that contains 1/SQRT(d).

Step 2: Create the plot in Figure 5.31. Right click on one of the data points. A pop-up menu will appear.

Step 3: Choose **Add Trendline...** from the pop-up menu. The Add Trendline dialog box will appear.

Grain Size	**Yield Strength**
d	σ_y
(μm)	(MN/m^2)
406	93
106	129
75	145
43	158
30	189
16	233

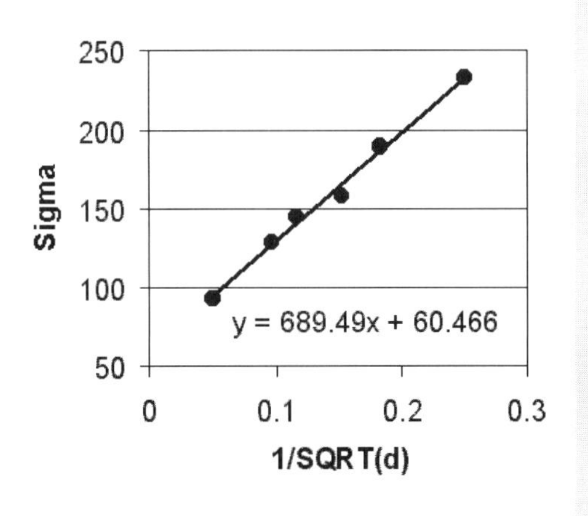

Figure 5.30. Table of carbon-steel data and results of regression analysis.

Step 4: Choose the **Type** tab and choose **Linear** from the box labeled Trend/Regression type.

Step 5: Choose the Options tab and check the box labeled *Display equation on chart*.

Step 6: Click **OK**.

Excel finds the least squares fit through the data points and displays the results on the graph, as shown in Figure 5.32.

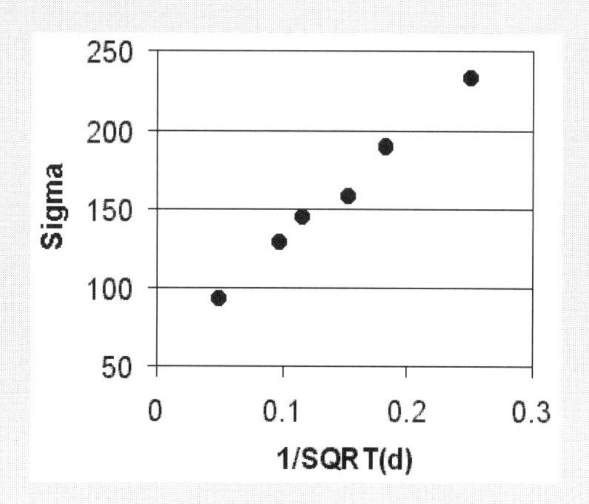

Figure 5.31. A plot of 1/SQRT(d) vs. σ_y.

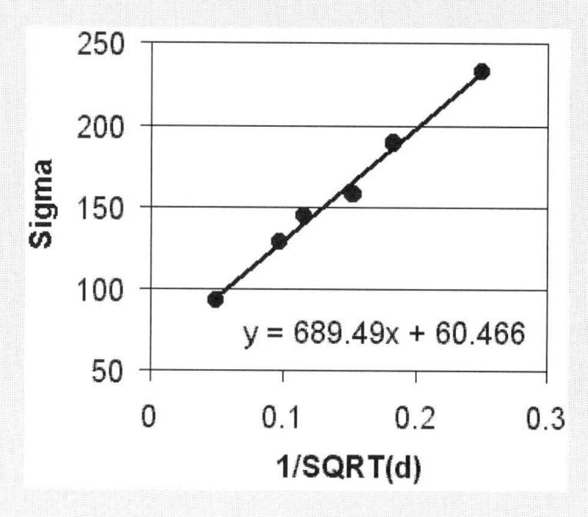

Figure 5.32. Plot with trendline and regression equation.

KEY TERMS

Analysis ToolPak	bin	constraints
correlation	descriptive statistics	exponential trend
Goal Seeker	histogram	linear extension
linear-regression analysis	linear replacement	linear trend
local minima	multiple regression	objective function
optimization	step value	stop value
trend analysis	trendline	

SUMMARY

This chapter introduced you to many of the powerful data-analysis tools offered by Excel. The tools presented in the chapter can be used to solve problems in your engineering courses. These include a large number of statistical and engineering functions in the Analysis ToolPak, a method for solving equations by using the Goal Seeker, and methods for solving linear and nonlinear optimization problems by using the Solver.

Problems

1. Use the Analysis ToolPak Histogram feature and the data in Figure 5.7 to create a cumulative percentage histogram of the students' midterm exam grades. What percent of the students earned a score of 80 or less on the midterm exam?
2. Use the Descriptive Statistics selection from the Analysis ToolPak to find the mean and standard deviation for the student GPAs in Figure 5.7. Compute how

many GPAs are within one standard deviation of the mean. Compute this by adding the standard deviation to the mean and then subtracting the standard deviation from the mean. How many GPAs lie with the range of the two numbers? What percentage of the GPAs lie within one standard deviation of the mean?

3. Look at the traffic-study data in Table 4.3. Perform a regression analysis for intersection 3. What is the linear equation that best fits the data? Use your linear equation to *predict* the traffic flow at this intersection in 2003.

4. Generate a scatter plot for the 11% column in Figure 2.22. Add two more X data points, for 50 years and 60 years. Create an exponential trendline for the curve. Choose the options box on the Format Trendline dialog box, and forecast forward 2 units (20 more years). By looking at the projected trendline, determine John's approximate nest egg at age 78.

5. Graph the function

$$f(x) = x^3 + \sin(x/2) + 2x - 4,$$

for $x = [-2, 2]$ in increments of 0.1. Add a trendline to the chart. What type of trendline best fits the data? Display the equation and R^2 value on the chart.

6. Generate data points for the function

$$y = \ln(2x) + \sin(x),$$

for $x = [1, 10]$ in increments of 0.1. Chart the results by using a line chart. From looking at the plot, what do you think is the minimum of the function in this range?

7. Compute the values of

$$f(x) = 2x^3 - 13x - 9,$$

for $x = [-3, 3]$ in increments of 0.1. Plot the function by using a line plot. You can see from the plot that $f(x) = 0$ near $x = -2.1, -0.8$, and 2.8. Use these three guesses and the Goal Seeker to find more accurate solutions for $f(x) = 0$.

8. Use the Solver to minimize the objective function

$$f(x) = (x_1 + 2x_2 - 7)^2 + (2x_1 + x_2 - 5)^2,$$

for the constraints $-10 \leq x_1, x_2 \leq 10$. What are the values of x_1 and x_2 when $f(x)$ is at its minimum?

9. A cylindrical chemical petroleum tank is to be built to hold 6.8 m^3 of hazardous waste. Your task is to design the tank in a cost-effective manner by minimizing its surface area. Ignore the thickness of the walls in your design. Recall that the surface area S of a right-angled cylinder is

$$S = 2\pi r^2 + 2\pi rh \text{ square meters},$$

and the volume V of a cylinder is

$$V = \pi r^2 h \text{ cubic meters}.$$

Use the Solver to minimize r and h.

6

Database Management within Excel

6.1 INTRODUCTION

Microsoft Excel implements a rudimentary *database management system*[1] by treating lists in a worksheet as database records. This is helpful for organizing, sorting, and searching through worksheets that contain many related items. You can import complete databases from external *database management systems (DBMS)* such as Microsoft Access, Oracle, dBase, Microsoft FoxPro, and text files. You can create structured queries by using Microsoft Query that will retrieve selected information from external sources.

If you require a relational DBMS, then you are encouraged to use another, more complete software application such as Microsoft Access. However, the database functions within Excel are adequate for many problems. An example of one way that an engineer might use this functionality is to import experimental data that have been stored in a relational DBMS in order to perform analysis on the data, using Excel's built-in functions.

SECTION

OBJECTIVES

After reading this chapter, you should be able to:

- Create a database within Excel
- Enter data into an Excel database
- Sort a database on one or more keys
- Use filters to query databases

[1]A collection of programs that enables you to store, extract, and modify information from a data base

	A	B	C	D	E
1	**Last Name**	**First Name**	**GPA**	**Department**	**Class**
2	Clinton	Willie	3.51	Electrical	Junior
3	Smith	Randolph	2.98	Chemical	Senior
4	Simpson	Susie	3.92	Electrical	Senior
5	Smith	Christine	2.78	Civil	Junior
6	Washington	Frank	3.41	Mechanical	Junior
7	Chavez	Linda	3.12	Chemical	Senior

Figure 6.1. Example of a student database.

6.1.1 Database Terminology

A database within Excel is sometimes called a *list*. These two terms will be treated synonymously here. A *database* can be thought of as an electronic file cabinet that contains a number of folders. Each folder contains similar information for different objects. For example, each folder might contain the information about a student at a college of engineering. The database is the collection of all student folders.

The data in each folder are organized in a similar fashion. For example, each folder might include a student's first name, last name, social security number, address, department, class, etc.

Using database terminology, each folder is called a *record*. Each data item is stored within a *field* or *row*. The title for each data item is called a *field name*.

Any region in an Excel worksheet can be defined as a database. Excel represents each record as a separate row. Each cell within the row is a field. The heading for each column is the field name.

Figure 6.1 depicts a small student database. Rows 2 through 7 each represent a student record. Each record has 5 fields. The field names are the column headings (e.g., *Last Name*).

6.2 CREATING DATABASES

Most database management systems store records in one or more separate files. The file delimits the boundaries of the database. Excel, however, stores a database as a region in a worksheet.

Excel must have some way of knowing where the database begins and ends in the worksheet. There are two methods for associating a region with a database. One method is to leave a perimeter of blank cells around the database region. The second is to explicitly name the region. Because of the unique way that Excel delimits a database, the following tips are recommended:

- Maintain only one database per worksheet. This will speed up access to the sorting and filtering functions, and you will not need to name the database regions.
- Make sure that each column heading in the database is unique. If there were two headings for *Last Name*, for example, a logical query such as:

  Find all records with Last Name equal to Smith

  would not make sense.

- Create an empty column to the right of the database and an empty row at the bottom of the database. Excel uses the empty row and column to mark the edge of the database. An alternate method is to assign a name to the region of the database. One disadvantage of assigning a name is that the allocated region may have to be redefined when records are added or deleted.
- Do not use cells to the right of the database for other purposes. Filtered rows may inadvertently hide these cells.

6.3 ENTERING DATA

Once the field names for the database have been created in the column headings, data may be entered by using several methods. One method of data entry is to directly type data into a cell. A database field may contain any legitimate Excel value, including numerical, date, text, or formula. For example, you might add a column to the database in Figure 6.1 that is titled *Full Name*. Instead of copying or retyping the first and last names of each student, the new field could concatenate the *First Name* and *Last Name* entries by using the following formula:

=CONCATENATE(B2," ",A2)

A second method for entering data is to use a form. To access the data entry form,

Step 1: Select any data cell in the database.

Step 2: Choose **Data → Form** from the Menu bar. The Data Entry form will appear, as depicted in Figure 6.2. The title of the Data Entry form will be the same as the name of the current worksheet.

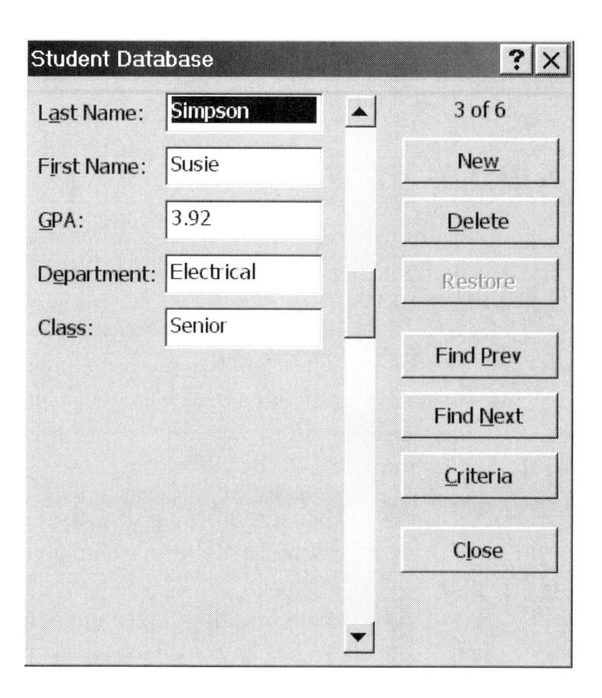

Figure 6.2. A Data Entry form.

From the Data Entry form, a new record can be created by clicking the **New** button. From this form, you can also scroll through the database, delete records, and modify existing records. The Data Entry form can also be used to filter data. This feature is explained in the next section.

PRACTICE!

Before proceeding, it will be helpful for you to create the database depicted in Figure 6.1. This database will be used for the examples in the rest of the chapter. Practice entering some of the data by using the Data Entry form. Enter some of the data by typing directly into the worksheet. Which method is more resistant to typing errors?

6.4 SORTING A DATABASE

The power of a database management system lies in its ability to search for information, rearrange data, and filter information.

To sort a database,

Step 1: Select any data cell in the database.

Step 2: Choose **Data** → **Sort** from the Menu bar. The Sort dialog box will appear, as depicted in Figure 6.3. The field on which the sort is made is called the *sort key*. Excel allows you to sort on multiple keys.

Step 3: Choose **Last Name** in ascending order as the first key. Choose **First Name** in ascending order as the second key.

Step 4: Make sure that the box labeled *Header Row* is checked.

Step 5: Click **OK**. The result is an alphabetical listing, by name, of the student database.

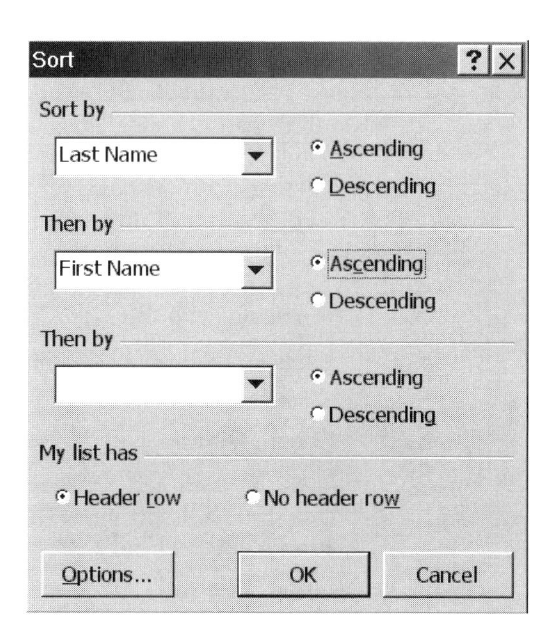

Figure 6.3. The Sort dialog box.

Warning: Be sure to select the entire database before sorting. If some columns are left out of the sort, then the database may become scrambled. For example, if you select only the *Last Name* and *GPA* columns and sort on GPA, the last names of the students will then become associated with the wrong first names!

If you accidentally scramble the database, immediately choose **Edit → UndoSort** from the Menu bar. The easiest way to select the entire database is to click on a single cell before performing any database operations. The entire database will then be highlighted.

PRACTICE!

Sort the student database in ascending order by Department and descending order of GPA within each department. Your results should resemble Figure 6.4.

	A	B	C	D	E
1	**Last Name**	**First Name**	**GPA**	**Department**	**Class**
2	Chavez	Linda	3.12	Chemical	Senior
3	Smith	Randolph	2.98	Chemical	Senior
4	Smith	Christine	2.78	Civil	Junior
5	Simpson	Susie	3.92	Electrical	Senior
6	Clinton	Willie	3.51	Electrical	Junior
7	Washington	Frank	3.41	Mechanical	Junior

Figure 6.4. Sorted student database (ascending Department and descending GPA).

6.5 SEARCHING AND FILTERING

Excel has several mechanisms for locating records that match specified criteria. For example, you may be interested in reviewing the students with Chemical Engineering majors. After you specify the criterion of department as Chemical, Excel displays only those records with Department field equal to 'Chemical'.

The process of limiting the visible records based on some criteria is called *filtering*. There are three general methods for filtering a database in Excel. The easiest methods are the use of the Data Entry form and the AutoFilter function. The Advance Filter function allows you to search by using more sophisticated logical criteria.

6.5.1 Filtering with the Data Entry Form

To use the Data Entry form to filter and search a database,

Step 1: Select a cell within the database.

Step 2: Choose **Data → Form** from the Menu bar. A Data Entry form will appear.

Step 3: Select the **Criteria** button on the Data Entry form. A blank record will appear.

Step 4: Type **Smith**, for example, in the Last Name field. Now repeatedly click the **Find Next** and **Find Prev** buttons. Note that only students with a last name of Smith appear.

Step 5: Choose **Criteria → Clear → Form** from the Data Entry form to clear the filter. Now, if you click the **Find Next** and **Find Prev** buttons, all of the students will appear.

6.5.2 Using the AutoFilter Function

The AutoFilter function allows you to filter records while viewing the database as a worksheet. To turn on the AutoFilter function,

Step 1: Select a cell within the database.

Step 2: Choose **Data → Filter → AutoFilter** from the Menu bar. A small arrow will appear in the heading of each column. When you click on one of the arrows, a small drop-down menu will appear that contains the possible choices for that field. Figure 6.5 depicts the student database with the AutoFilter option turned on. The drop-down menu for the *Department* field is shown.

Step 3: Choose a department from the menu (e.g., **Civil**). Only the records with Department equal to *Civil* will be displayed. The small arrow at the head of the *Department* column will change color to signify that this column is filtering some records.

Step 4: Choose **Data → Filter → Show All** from the Menu bar to redisplay all records.

PRACTICE!

Practice using the AutoFilter. Use the AutoFilter to select all seniors in Electrical Engineering.

6.5.3 Using the Custom AutoFilter

The AutoFilter options that you have learned so far are fine if you want to make an exact match. However, in many cases, you will want to specify a range. For example, you may want to view the students whose GPA is greater than 3.0. One way to specify ranges is to use the Custom AutoFilter dialog box. To use the Custom AutoFilter option to specify students with a GPA > 3.0,

Step 1: Select a cell within the database.

Step 2: Choose **Data → Filter → AutoFilter** from the Menu bar. Choose the AutoFilter arrow for the *GPA* column.

Step 3: Choose **(Custom . . .)** from the drop-down menu. The Custom AutoFilter dialog box will appear, as shown in Figure 6.6.

Step 4: Choose **is greater than** from the menu labeled *GPA*.

Step 5: Type **3.0** into the right-hand box under *GPA*. This is shown in Figure 6.6.

Step 6: Click **OK**.

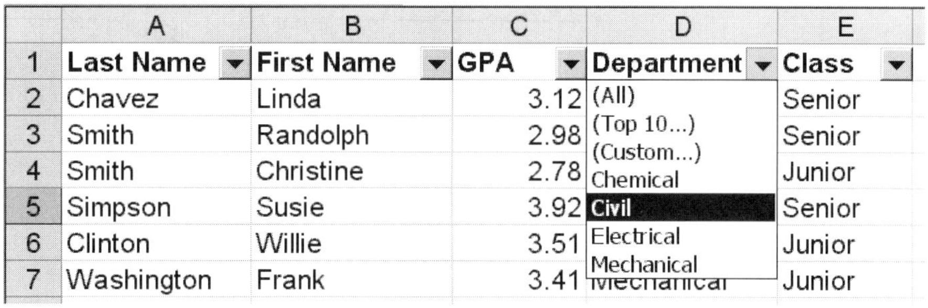

	A	B	C	D	E
1	Last Name	First Name	GPA	Department	Class
2	Chavez	Linda	3.12	(All)	Senior
3	Smith	Randolph	2.98	(Top 10...) (Custom...)	Senior
4	Smith	Christine	2.78	Chemical	Junior
5	Simpson	Susie	3.92	Civil	Senior
6	Clinton	Willie	3.51	Electrical	Junior
7	Washington	Frank	3.41	Mechanical	Junior

Figure 6.5. A sample AutoFilter menu.

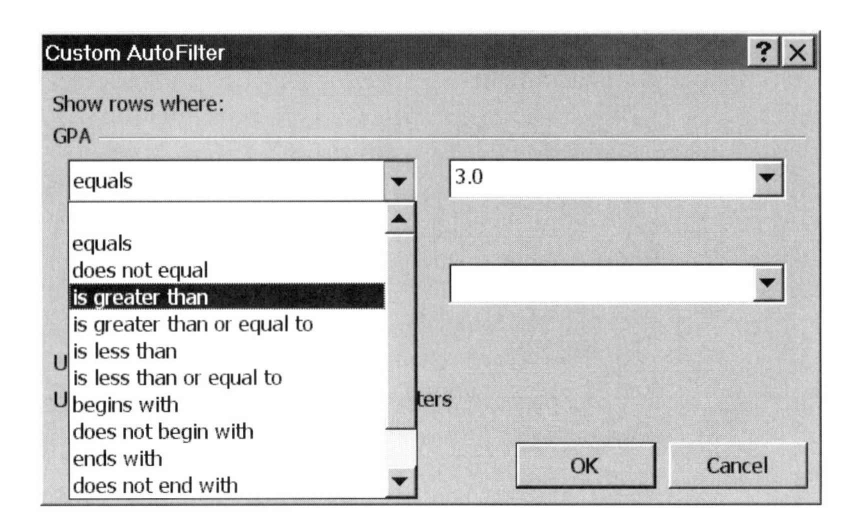

Figure 6.6. The Custom AutoFilter dialog box.

PRACTICE!

Practice using the Custom AutoFilter function. You can use it to specify a simple logical expression. The Custom AutoFilter dialog box allows you to join two logical conditions with an *And* operator or an *Or* operator.

If you choose **And**, then both conditions must be true for the record to be displayed. If you choose **Or**, then the record will be displayed if either of the conditions is met.

In Figure 6.7, the Custom AutoFilter dialog box is used to select all students with GPA ≤ 3.5 and GPA ≥ 3.0.

Clear the filter by choosing **Data → Filter → ShowAll** from the Menu bar. Then, see if you can display all students from either Electrical or Chemical Engineering. Use the *Or* operator for this task.

Custom AutoFilter

Show rows where:

GPA

is less than or equal to 3.5

And Or

is greater than or equal to 3.0

Use ? to represent any single character

Use * to represent any series of characters

OK Cancel

Figure 6.7. Using the *And* operator.

6.5.4 Using Wild-card Characters

A greater range of choices may be made by using the question mark (?) or asterisk (*) characters as wild-card characters. A *wild-card character* is used as a placeholder that can be filled by any legitimate character.

The question mark is used to represent the replacement of any single character. For example, the logical expression

Department = ????ical

will return all departments with exactly four characters followed by *ical*. In our student database, Chemical would be displayed, but Mechanical, Electrical, and Civil would not, as they do not contain exactly 4 characters followed by *ical*.

The asterisk is used to represent the replacement of zero or more characters. For example, the logical expression

Department = *ical

will return all departments ending in *ical*, no matter how many letters precede *ical* (including zero letters). In our student database, Chemical, Electrical, and Mechanical would be displayed, but Civil would not.

Some wild-card functions can also be performed by using the following conditions in the Custom AutoFilter dialog:

- Begins with
- Ends with
- Contains
- Does not begin with
- Does not end with
- Does not contain

PRACTICE!

Practice using the wild-card characters and menu selections in the Custom AutoFilter dialog box. Be sure to clear the filter between each exercise.

Here are two methods of selecting all students whose last name begins with an *S*:

Method 1:

Last Name begins with S

Method 2:

Last Name equals S*

Here are two methods for selecting all students whose last name ends in an *N* and does not contain a *W*:

Method 1:

Last Name ends with N
And
Last Name does not contain W

Method 2:

Last Name equals *N
And
Last Name does not equal *W*

Note that the Custom AutoFilter box is not case sensitive.

6.5.5 Using the Advanced Filter Function

The filtering methods that you have been shown so far allow a great deal of flexibility. By adding filters to multiple fields and by using the Custom AutoFilter function, wild-cards, and logical operators, you can build relatively complex filters.

However, there are some cases that can't be handled by these methods. Suppose that the engineering departments have different GPA requirements. We want to find students who meet the following criteria:

Electrical Engineering with GPA > 3.6
Or
Chemical Engineering with GPA > 3.0
Or
Civil Engineering with GPA > 3.2
Or
Mechanical Engineering with GPA > 3.4

You can't create a filter that solves this request by using the AutoFilter functions. You will need to use the Advanced Filter function, which allows you to build more complex queries.

To use the Advanced Filter function, you must first set up a Criteria table. A *Criteria table* is just what it sounds like, a table of criteria that must be met for a filter to occur. To create a Criteria table,

Step 1: Copy the field names from your database to another location in the same worksheet.
Step 2: Leave at least one blank row of cells between the Criteria table and the database.
Step 3: Type in the criteria that must be met for your filter. See the example in Figure 6.8.

6.5.6 Logic within Rows

All criteria within a single row must be met for a match to occur. This is equivalent to a logical *And* operator.

6.5.7 Logic between Rows

A Criteria table may have more than one active row. A match on any row in a Criteria table may be met for a match to occur. This is equivalent to a logical *Or* operator. Figure 6.8 depicts a Criteria table and our example student database.

	A	B	C	D	E
1	**Criteria Table**				
2	**Last Name**	**First Name**	**GPA**	**Department**	**Class**
3			>3.6	Electrical	
4			>3.0	Chemical	
5			>3.2	Civil	
6			>3.4	Mechanical	
7					
8	**Database**				
9	**Last Name**	**First Name**	**GPA**	**Department**	**Class**
10	Chavez	Linda	3.12	Chemical	Senior
11	Smith	Randolph	2.98	Chemical	Senior
12	Smith	Christine	2.78	Civil	Junior
13	Simpson	Susie	3.92	Electrical	Senior
14	Clinton	Willie	3.51	Electrical	Junior
15	Washington	Frank	3.41	Mechanical	Junior

Figure 6.8. Example of a Criteria table.

Now that we have set up the Criteria table, we can use the Advanced Filter function to build a filter by using the Criteria table. Figure 6.8 shows the criteria that solve our earlier problem of finding engineering students who would meet the different GPA requirements of the engineering department within different fields of engineering. You can see from a visual inspection that three students meet our requirements: Susie, Linda, and Frank. To use the Advanced Filter function,

Step 1: Choose **Data → Filter → Advanced Filter** from the Menu bar. The Advanced Filter dialog box will appear, as shown in Figure 6.9.

Step 2: Select the database range in the box titled *List range*. Include the field names in your selection.

Step 3: Select the Criteria table in the box titled *Criteria range*. Include the field names in your selection. You can filter the list in place, or you can have the filtered results displayed in another location.

Step 4: Check the box titled *Filter the list, in-place*.

Step 5: Click **OK**. Your results should resemble Figure 6.10.

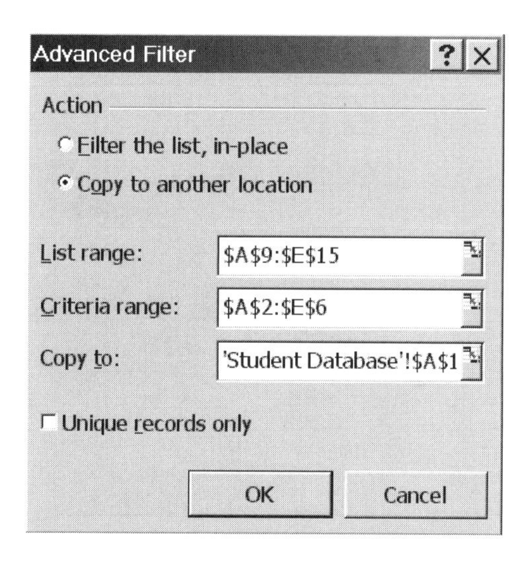

Figure 6.9. The Advanced Filter dialog box.

	A	B	C	D	E
1	Criteria Table				
2	Last Name	First Name	GPA	Department	Class
3			>3.6	Electrical	
4			>3.0	Chemical	
5			>3.2	Civil	
6			>3.4	Mechanical	
7					
8	Database				
9	Last Name	First Name	GPA	Department	Class
10	Chavez	Linda	3.12	Chemical	Senior
13	Simpson	Susie	3.92	Electrical	Senior
15	Washington	Frank	3.41	Mechanical	Junior

Figure 6.10. Results from the Advanced Filter example.

KEY TERMS

And operator	Criteria table	database
database management system	DBMS	field
field name	filtering	list
Or operator	record	row
sort key	wild-card characters	

SUMMARY

In this chapter, you were shown how to access and use the Excel database features. These include the ability to perform multilevel sorts. The search and filter tools were described, including the AutoFilter and the Advanced Filter functions.

Problems

Figure 6.11 shows the thermal properties of sundry materials. The construction column indicates whether the material is used in typical residential construction. Type this data into a worksheet, or download the worksheet from the author's Web site:

http://www.pobox.com/~kuncicky/Excel2002

	A	B	C	D	E
1	Thermal Properties of Various Materials (at 20° C)				
2					
3	Material	Construction	Density - ρ (kg/m^3)	Specific Heat - c_p (J/kg·K)	Conductivity - *k* (J/s·m·°C)
4	Aluminum	yes	2700	896	237.000
5	Bronze	no	8670	343	26.000
6	Concrete	yes	500	840	0.130
7	Copper	no	8930	383	400.000
8	Glass	yes	2800	800	0.810
9	Ice	no	910	57	2.200
10	Plaster	yes	1800	112	0.810
11	Polystyrene	yes	1210		0.040
12	Wood (pine)	yes	420	2700	0.150
13	Wool insulation	yes	200		0.038

Figure 6.11. Thermal properties of various materials.

From the Web page, select the item labeled *Thermal Properties*. Use this worksheet to solve the following problems:

Step 1: Sort the worksheet by specific heat in ascending order. Where do the blank entries get placed after the sort?

Step 2: Sort the worksheet by the *Construction* field with the *yes* category at the top. Within each construction category, sort in descending order of density.

Step 3: Use the AutoFilter function to show only construction materials.

Step 4: Use the Custom option of the AutoFilter function to show only construction materials with a density > 1000 and conductivity < 1.00.

Step 5: Use the Advanced Filter function to show construction materials with a specific heat > 800 and nonconstruction materials with a specific heat > 300. Show your Criteria table.

Step 6: Can you solve Problem 5, using only the AutoFilter, not the Advanced Filter?

Describe the actions you performed to solve each problem.

7

Collaborating with Other Engineers

7.1 THE COLLABORATIVE DESIGN PROCESS

Engineering design is the process of devising an effective, efficient solution to a problem. The solution may take the form of a component, a system, or a process. Engineers generally solve problems by collaborating with others as members of a team. As a student, you will undoubtedly be asked to participate in collaborative projects with other students.

You may or may not have experience working on a team. If a team works together effectively, then more can be accomplished by the team than through any individual effort (or even the sum of individual efforts). If team members do not work together effectively, however, the group can become mired in power struggles and dissension. When this occurs, one of two things usually happens. Either the team makes little progress towards its goals, or a small subgroup of the team takes charge and does all of the work. Some guidelines for being an effective team member are presented at the end of this chapter, in the "Professional Success" section.

SECTION

OBJECTIVES

After reading this chapter, you should be able to:

- Track revisions in an Excel document
- Share workbooks among team members
- Insert comments in an Excel document
- Transfer worksheet data to and from other applications
- Use a password to restrict ability to open a file
- Use a password to restrict ability to write to a file
- Use a password to restrict access to a worksheet

7.1.1 Microsoft Excel and Collaboration

The ability to work well on a team can best be learned by participating on a successful team. Microsoft Excel includes several tools that can help to solve one of the most burdensome technical tasks of group collaboration—the preparation of the team document. In the past, collaborative document preparation has been extremely difficult. The result has been that the task is usually assigned to one or two team members. New features of Microsoft Excel make it feasible for the whole team to participate in the composition and revision of a document. Learning to use these features will require some time and practice on your team's part. The rewards, however, will be well worth the effort.

7.2 TRACKING CHANGES

One problem that arises in document preparation by a team is keeping track of revisions. For example, one team member may be given the task of revising a portion of the team project. After the revisions are made, the team will meet and approve some, or all, of the revisions. Then one of the team members will incorporate the accepted changes into the document.

Excel has a feature called *Tracking Changes* that will not only mark revisions, but will also keep track of who is making each revision. The worksheet may be printed showing both the original text and the new revisions. Revisions may then be globally accepted or selectively accepted into the document.

7.2.1 Highlighting Changes

To turn on the tracking changes feature,

Step 1: Choose **Tools → Track Changes → Highlight Changes** from the Menu bar. The Highlight Changes dialog box will appear, as shown in Figure 7.1.

Step 2: Check the box labeled *Track changes while editing*. This will let you share the workbook, and it will turn on history tracking. You will now have access to the next three boxes and drop-down lists, which allow you to limit the changes that are highlighted by time, user, and worksheet region.

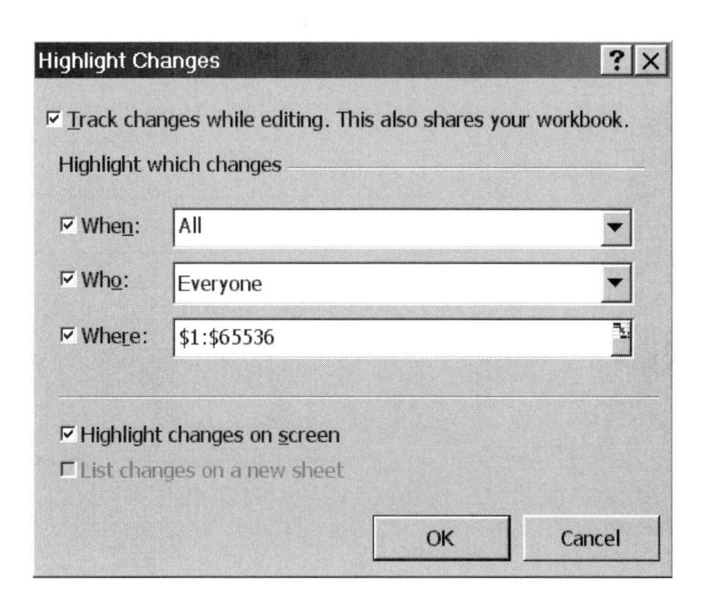

Figure 7.1. The Highlight Changes dialog box.

Step 3: Check the box labeled *When* and choose **All**.

Step 4: Check the box labeled *Who* and choose **Everyone**.

Step 5: Check the box labeled *Where* and select the entire worksheet.

Step 6: Check the box labeled *Highlight changes on screen*, which is located at the bottom left of the dialog box and is one of the two items that let you decide where to display the revisions. Any revisions made to the current worksheet will now be highlighted on the screen. (See Figure 7.2.) The alternate selection saves revisions to a new worksheet.

Step 7: Type a few words in your worksheet and notice what happens. Any modified cells are outlined in blue, and a small blue tab is placed in the upper left corner of the cell. These are called *revision marks* (See Figure 7.2.)

7.2.2 Creating an Identity

As you review a document that other team members have revised, you can see who made the revision, the date and time of the revision, and the previous contents of the cell. The identity feature works only if each reviewer has given an identity to the Excel application. To identify yourself to Excel,

Step 1: Choose **Tools → Options** from the Menu bar. The Options dialog box will appear.

Step 2: Select the tab labeled **General**.

Step 3: Type your name in the box labeled *User Name*.

Your identity will now be attached to any revisions that you make to a worksheet. You can test the feature by making a few revisions and then moving the mouse cursor over a region that has revision marks. A small box will appear that displays the reviewer's name along with a date and time stamp. An example is shown in Figure 7.3.

	A	B
1	Height (in)	Weight (lb)
2	62	135
3	74	190
4	68	155
5	71	210

Figure 7.2. Examples of highlighted, revised cells.

	A	B	C	D	E	F
1	Height (in)	Weight (lb)				
2	62	135				
3	74	190				
4	68	155				
5	71	210				
6						
7						
8						

Dave Kuncicky, 3/8/2002 11:08 AM:
Changed cell B5 from '<blank>' to '210'.

Figure 7.3. Example revision marks with reviewer's information.

7.2.3 Incorporating or Rejecting Revisions

Revision marks are not wholly incorporated into the document until they are reviewed and then accepted or rejected. Revision marks can be reviewed by using the Accept or Reject Changes dialog box. To accept or reject changes,

Step 1: Choose **Tools → Track Changes → Accept or Reject Changes** from the Menu bar. The Select Changes to Accept or Reject dialog box will appear, as shown in Figure 7.4. You will now have access to the three boxes and drop-down lists, allowing you to select which changes to review.

Step 2: Check the box labeled *When* and choose **Not yet reviewed** from the drop-down menu. This will allow you to review all changes that have not yet been reviewed.

Step 3: Check the box labeled *Who* to choose whose changes to review. Select **Everyone**.

Step 4: Check the box labeled *Where* and select the entire worksheet.

Step 5: Click **OK**, when you are finished with your selections.

Excel will now guide you through each selected revision and give you the opportunity to accept or reject the revision. The Accept or Reject Changes dialog box will appear, as shown in Figure 7.5. You will be guided through the revisions one at a time, unless you select **Accept All** or **Reject All**.

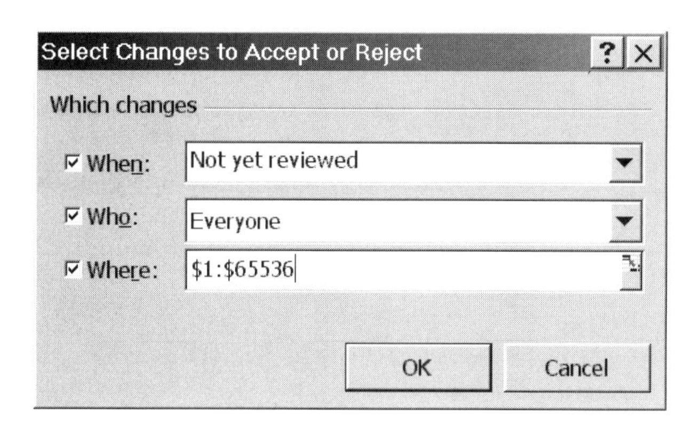

Figure 7.4. Select Changes to Accept or Reject dialog box.

Accept or Reject Changes
Change 1 of 3 made to this document:
Dave Kuncicky, 3/8/2002 11:12 AM:
Auto-Inserted row 5.
Choosing 'Reject' will also reject the following:
Changed cell A5 from '<blank>' to '71'.
Accept | Reject | Accept All | Reject All | Close

Figure 7.5. The Accept or Reject Changes dialog box.

7.3 ADDING COMMENTS TO A DOCUMENT

At times, a reviewer may want to attach notes or *comments* to a cell without changing the contents of the cell. To add a comment to a cell,

Step 1: Select the cell to comment.

Step 2: Choose **Insert → Comment** from the Menu bar. A Comment box will appear, as shown in Figure 7.6.

Step 3: Type in a comment and then click outside of the Comment box to exit.

Step 4: A cell that is attached to a comment will be marked with a small red tab in the upper right-hand corner of the cell. When the cell is selected, the comment will reappear.

Comments may be reviewed, edited, or deleted by two methods. To review a single comment,

Step 1: Select the cell holding the comment and then click the right mouse button. The drop-down Quick Edit menu will display several new items, as shown in Figure 7.7.

Step 2: Use one of the options from the Quick Edit menu: view, modify, or delete the selected comment.

If there are many comments to review, then the preceding method becomes cumbersome. If you are going to review a large number of comments, then the Reviewing toolbar is easier to use. To use the Reviewing toolbar,

Step 1: Choose **View → Toolbars** from the Menu bar, and make sure that the box labeled *Reviewing* is checked. The Reviewing toolbar, which is displayed in Figure 7.8 contains icons for moving through the comments in a worksheet one at a time and selectively viewing, editing, or deleting each comment.

Step 2: Select one of the buttons that best describes your needs. The buttons related to comments are described in Table 7.1. You can use the buttons to quickly review and modify the comments.

	A	B	C	D
1	Height (in)	Weight (lb)	**Dave Kuncicky:** Should we be using metric units here?	
2	62	135		
3	74	190		
4	68	155		
5	71	210		

Figure 7.6. Example of a Comment box.

Figure 7.7. Comment-related items on the Quick Edit menu.

Figure 7.8. The Reviewing toolbar.

TABLE 7.1 The Icons on the Reviewing Toolbar

ICON	Name	ACTION
	New Comment	Creates a new comment for the selected cell.
	Previous Comment	Moves to the previous comment in the worksheet.
	Next Comment	Moves to the next comment in the worksheet.
	Show/Hide All Comments	Toggles between showing and hiding all comments.
	Delete Comment	Deletes the current comment.
	Send E-mail	Opens an e-mail dialog for sending the current worksheet as an attachment.

7.4 MAINTAINING SHARED WORKBOOKS

Excel provides a mechanism that allows several users to simultaneously share a workbook over a network. A shared workbook must reside in a shared folder on the network. Other access restrictions may apply, depending on your local network setup.

Once you are able to share a workbook, different users can view and modify the workbook at the same time. Sharing a document clearly requires some protocol among the group in order to keep several users from overwriting each other's work. Sharing a workbook is most effective if simultaneous users edit different parts of the workbook. Excel can be set to keep a history of changes to a shared workbook, and previous versions may be recalled if necessary.

7.4.1 Sharing a Workbook

To share a workbook,

Step 1: Choose **Tools → Share Workbook** from the Menu bar. The Share Workbook dialog box will appear, as shown in Figure 7.9.

Step 2: Select the **Editing** tab.

Step 3: Check the box labeled *Allow changes ... etc.*, to allow the workbook to be shared. Once the workbook is shared, this tab can be used to see who is currently editing the workbook.

7.4.2 Keeping a Change History

Excel can keep a log of changes made by each user of a shared workbook. The log of changes is called a *change history*. To set the options for a change history,

Step 1: Choose **Tools → Share Workbook** from the Menu bar. The Share Workbook dialog box will appear, as shown in Figure 7.9.

Step 2: Select the **Advanced** tab. The Advanced screen of the Share Workbook dialog box will appear, as shown in Figure 7.10.

Step 3: In the section labeled *Track changes*, set the length of time to keep a change history.

Step 4: Check the box titled *Don't keep change history* to turn the change history off. One reason to turn off the change history or to keep the time duration low is to limit the size of the workbook. A change history can significantly increase the disk space required to store a workbook. There is a trade-off between caution and storage requirements. The use of the change-history feature is not a substitute for regularly backing up a workbook to some other medium, such as a removable disk or tape.

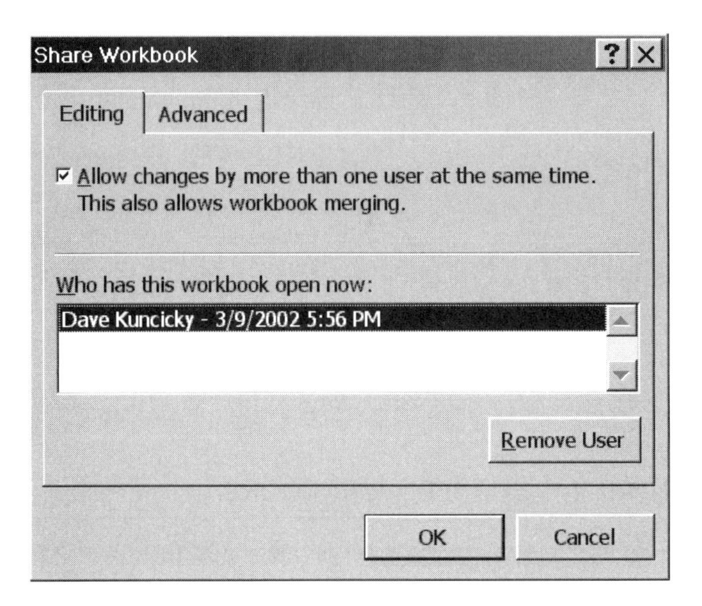

Figure 7.9. The Editing tab on the Share Workbook dialog box.

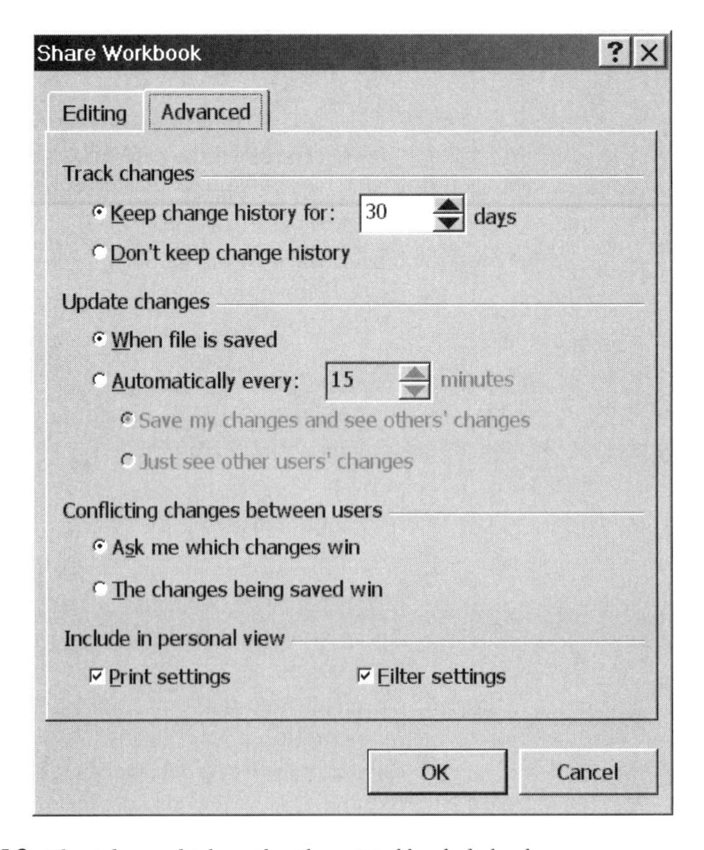

Figure 7.10. The Advanced tab on the Share Workbook dialog box.

7.4.3 Timed updates

The second section of the Shared Workbook dialog box's Advanced Setting tab is used to specify when changes are updated so that other users may see them. The first selection specifies that your changes will be updated to the group whenever you save the file. Alternatively, you can choose to have the changes automatically update the other users' view of the workbook every few minutes.

7.4.4 Managing Conflicts

If you are about to save a workbook, some of your changes may conflict with pending changes from another user. The third section of the Advanced Setting tab allows you to specify how you want to resolve conflicts, if at all. If you choose the first option titled *Ask me which changes win*, then the Resolve Conflicts dialog box will appear when you save the file.

You will be prompted to resolve each conflict. If you don't want to resolve conflicts when you save a shared workbook, then click the item titled *The changes being saved win*. The last user to save conflicting changes wins.

7.4.5 Personal Views

The last section of the Advanced Setting tab allows the creation of personal printer or filter settings. When the workbook is saved, a separate personal view is saved for each user.

7.4.6 Merging Workbooks

Group members do not always have access to the same network. One scenario that occurs when groups collaborate on a workbook is that each member takes a copy of the workbook. Each group member works separately on the workbook, and later the workbooks are merged into a single document.

Copies of a workbook can be revised and merged only if a change history is being maintained. Be sure to set a sufficient length of time for the change history so that the history doesn't expire before the workbook copies are merged. The number of days is set in the Share Workbook dialog box.

To merge several copies of a workbook,

Step 1: Open the first copy and then choose **Tools → Compare and Merge Workbooks** from the Menu bar. You will be prompted to choose a file to merge.

Step 2: Continue to merge files until all of the copies have been merged into one workbook.

Follow the instructions in the next section to view the history of all of the changes that have been made.

7.4.7 Viewing the History of Changes

After the workbooks have been merged, you may want to view the change history and selectively choose the changes that you want in your updated version.

To accept or reject the charges,

Step 1: Choose **Tools → Track Changes → Accept or Reject Changes** from the Menu bar. The Accept or Reject changes dialog box will appear.

Step 2: Follow the instructions as you are guided through each change. The example in Figure 7.11 shows that cell B5 has been changed a number of times by two different users.

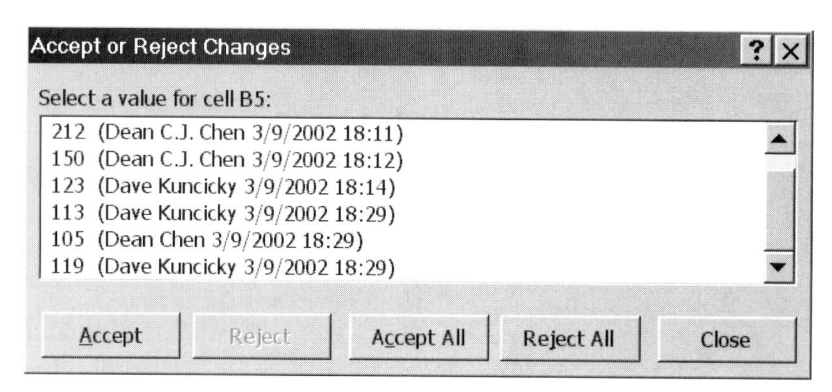

Figure 7.11. Reviewing a change history.

7.4.8 Restrictions for Shared Workbooks

Some features of Excel cannot be used while a workbook is being shared. However, all features can be used if the workbook has sharing turned off. The disadvantage of turning off sharing is that the change history is deleted. When a workbook is shared, a good rule of thumb is to use good judgment, as not all worksheets or workbooks are meant to be shared. Here's a list of some of the features of Excel that cannot be used when sharing is in effect:

- Creation, modification, or deletion of passwords. Passwords should be set up before the workbook is shared.
- Deletion of worksheets.
- Insertion or modification of charts, pictures, or hyperlinks.
- Insertion or deletion of regions of cells. (Single rows or columns can be deleted.)
- Creation of data-table or pivot-table.
- Insertion of automatic subtotals.

7.5 PASSWORD PROTECTION

Several levels of protection exist for workbooks. Your personal file space may be protected by the network operating system. The folder in which your workbook resides may be protected. These methods are outside the scope of this section. The methods that are discussed here apply only to a single workbook or parts of a workbook.

The methods below, in and of themselves, will not prevent another user from copying your workbook, but one of the methods (open access) can prevent another user from viewing the copied workbook.

One way to limit access to a shared workbook is with password protection. A variety of password types will be discussed next. In every case, be sure to write down or memorize your password. If you lose a password, you will not be able to retrieve your work.

7.5.1 Open Protection

A password can be set that restricts a user from opening a file. This means that an unauthorized user cannot read or print the file using Excel. This type of access is called *open access*, since it protects a file from being opened. A user will still be able to copy the file. The password protection for open access will also apply to the copied file. To set a password for open access,

Step 1: Choose **File → Save As.** The Save As dialog box will appear.

Step 2: Choose **Tools → General Options** from the Save As dialog box. The Save Options dialog box will appear, as shown in Figure 7.12.

Step 3: Choose the first option, *Always create backup*. This option specifies that Excel should create a backup copy of your workbook every time it is saved. Unless you are extremely short of disk space, this is an excellent option!

Step 4: To restrict others from opening your workbook file, type a password in the box titled *Password to open*. You will be prompted to type the password a second time for verification. Note that Excel uses case-sensitive passwords.

Hint: One of the most common reasons that a password seems to suddenly stop working is that you have the caps-lock key turned on.

7.5.2 Write Protection

There may be times where you want to allow read access to others, but you do not want anyone to be able to modify your original file. This type of protection is called *write access*. To set a write access password,

Step 1: Choose **File → Save As**. The Save As dialog box will appear.

Step 2: Choose **Tools → General Options** from the Save as dialog box. The Save Options dialog box will appear, as shown in Figure 7.12.

Step 3: Type a password in the box titled *Password to modify*. You will be prompted to type the password a second time for verification.

The next time you attempt to open the file, the Password dialog box will appear, as depicted in Figure 7.13. You will be prompted for a password if you want to open the file for write access. A password is not needed to open the file for reading only.

Note: A user can open a write-protected file as a read-only file and then save it under a different name. The new file can be modified by the user without a password!

7.5.3 Sheet Protection

Protection can be finely tuned. Once you have completed part of a worksheet, you may want to protect it merely to prevent yourself from inadvertently modifying that section. One example of the use of sheet protection is to lock cells that contain formulas, while allowing cells that contain data to be modified.

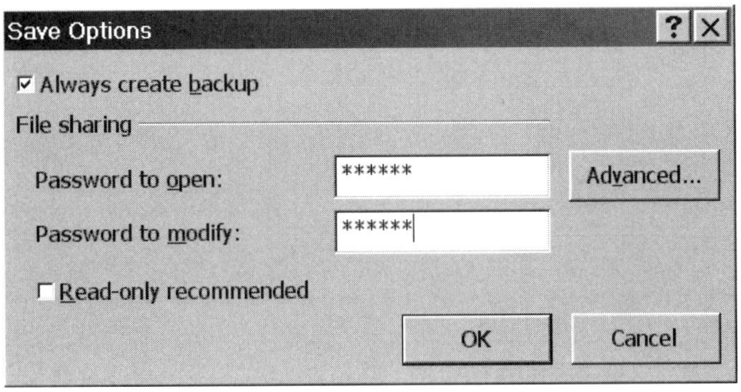

Figure 7.12. The Save Options dialog box.

Figure 7.13. The Password dialog box.

Before turning on sheet protection, you must lock the region of cells to be protected. For example, Figure 7.14 shows a worksheet with the quadratic formula. The formulas for the two solutions are in cells D2 and E2. The user inputs the three coefficients a, b, and c in cells A2, B2, and C2, respectively. We would like to lock the cells containing the formulas so that they can't be inadvertently modified.

To protect a region of cells within a workbook,

Step 1: As we want to lock the cells containing formulas, select the region D2:E2 and then choose **Format → Cells**. The Format Cells dialog box will appear.

Step 2: Select the **Protection** tab. The Protection page of the Format Cells dialog box is shown in Figure 7.15.

Step 3: Check the box labeled *Locked*, and uncheck the box labeled *Hidden*.

Step 4: Select the region A2:C2 to unlock the data-entry cells. Choose **Format → Cells**. The Format Cells dialog box will appear.

Step 5: Select the **Protection** tab.

Step 6: Uncheck both boxes labeled *Locked* and *Hidden*.

Locking cells alone has no effect. Before the locked cells are activated, you must turn on sheet protection. To turn on sheet protection,

Step 1: Choose **Tools → Protection → Protect Sheet** from the Menu bar. The Protect Sheet dialog box will appear as shown in Figure 7.16.

Step 2: Check the box titled *Protect worksheet and contents of locked cells*, which is one many options offered.

Step 3: Check the box lableled *Select unlocked cells*.

Step 4: Uncheck all other boxes.

Step 5: Type in a password if you wish. (If you do, you will be prompted to verify the password.)

Step 6: Click **OK**.

Pos_Root = (-B2 + SQRT(B2^2 - 4 * A2 * C2)) / (2 * A2)

	A	B	C	D	E
1	Coefficient A	Coefficient B	Coefficient C	Root 1	Root 2
2	2	-4	1	1.7071	0.2929

Figure 7.14. Solution for quadratic equations.

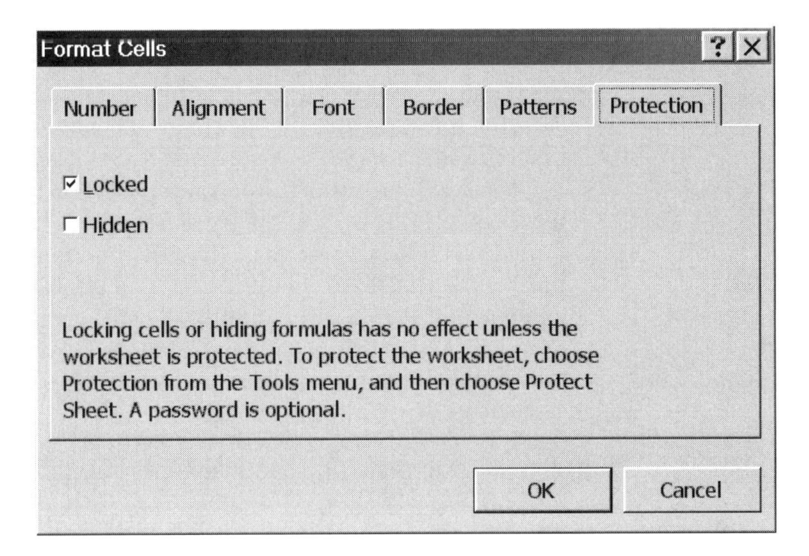

Figure 7.15. The Protection page of the Format Cells dialog box.

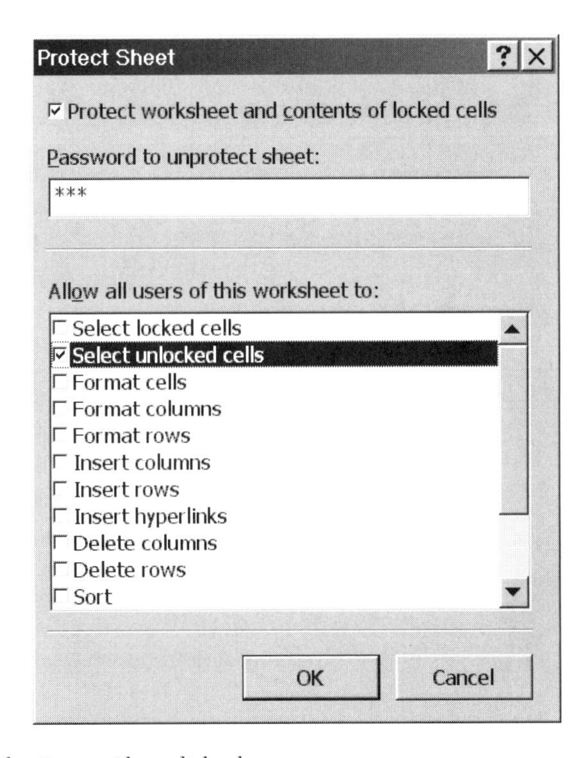

Figure 7.16. The Protect Sheet dialog box.

Note that you can now enter values for the coefficients into the quadratic formula worksheet, and the results will change. However, you cannot select cells D2 or E2 and, you cannot modify the formulas!

7.6 IMPORTING AND EXPORTING DATA FROM EXTERNAL FILE FORMATS

One problem of working as part of a team is that team members may use different application software. In addition, a large and complex project may require the use of several software packages, such as Word, Access, and HTML documents. This implies that you may have to move data from one application to another in the process of completing a project.

Excel provides several methods for importing data. The Microsoft Query program enables the selective retrieval of data from external database files such as Oracle, dBase, and Paradox. Web queries are methods that are designed to retrieve data from sites on the World Wide Web, using your default Web browser as an interface.

In this section, we will discuss two methods for importing external files: File Open method and the Text Import Wizard.

7.6.1 Import Using the File Open Option

We will now discuss the method for directly importing and exporting files to and from other file formats. A number of types of file formats may be imported directly into Excel. To import files,

Step 1: Choose **File → Open** from the Menu bar. The Open dialog box will appear.

Step 2: Click on the arrow to the right of the box labeled *Files of type*. A small drop-down menu will appear, as depicted in Figure 7.17.

A few of the importable file types are shown in Figure 7.17. These include Microsoft Access, Quattro Pro, dBase, etc.

An external file type may be opened and viewed within Excel. If the file is modified, Excel will ask if you want to save the file in its original format or in Excel format. If the file is saved in Excel format, then the external program (e.g., dBase) will not be able to view the changes. If the file is saved in the external format (e.g., dBase), then some Excel formatting may be lost. For example, formulas and macros may not be translated into the external format.

The methods for importing from and exporting to applications *not* listed on the File Open menu are specific to each brand of application software. For example, MathCAD provides data input and output filters for Excel, Lotus® 1-2-3, and MATLAB™. To seamlessly transfer data between Excel and MATLAB™ a separate piece of software called Excel Link must be purchased. Excel Link provides a means for exchanging data between MATLAB™ and Excel, taking advantage of Excel's familiar spreadsheet interface and the computational and visualization capabilities of MATLAB™.

7.6.2 Importing Text Data by Using the Text Import Wizard

Most applications will let you export data (using the *Save as* menu item) as tab- or space-delimited text. In addition, you may produce data from a computer program that you

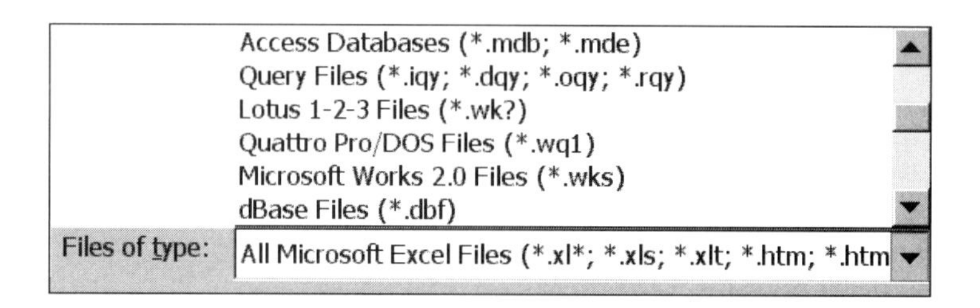

Figure 7.17. The Files of Type drop-down menu.

have written. In either case, the Text Import Wizard helps you align and import the data. Note that formatting, macros, colors, font size, etc., cannot be imported by using this method. Only data can be easily imported from text files.

To see an example of importing text,

Step 1: Use your favorite text editor (e.g., Notepad, WordPad, or Word) to create a tab-delimited file as displayed in Figure 7.18. Be sure to place a single tab between each column.

Step 2: Choose **Save As** from the Menu bar and save the file as type **Text Only** if you are using Microsoft Word or WordPad.

Step 3: Close the file, and choose **File → Close** from the Menu bar.

Step 4: Open the same file from Excel by choosing **File → Open** from the Menu bar.

Step 5: Choose **All Files** (*.*) from the list labeled *Files of Type*.

Step 6: Locate the file that you created and choose **Open**. Step 1 of the Text Import Wizard will appear, as depicted in Figure 7.19. Your data should now be visible in the Wizard dialog box.

```
7.1   9.3   10.4
2.5   21.0  13.0
3.4   12.2  98.4
```

Figure 7.18. Tab-delimited text data.

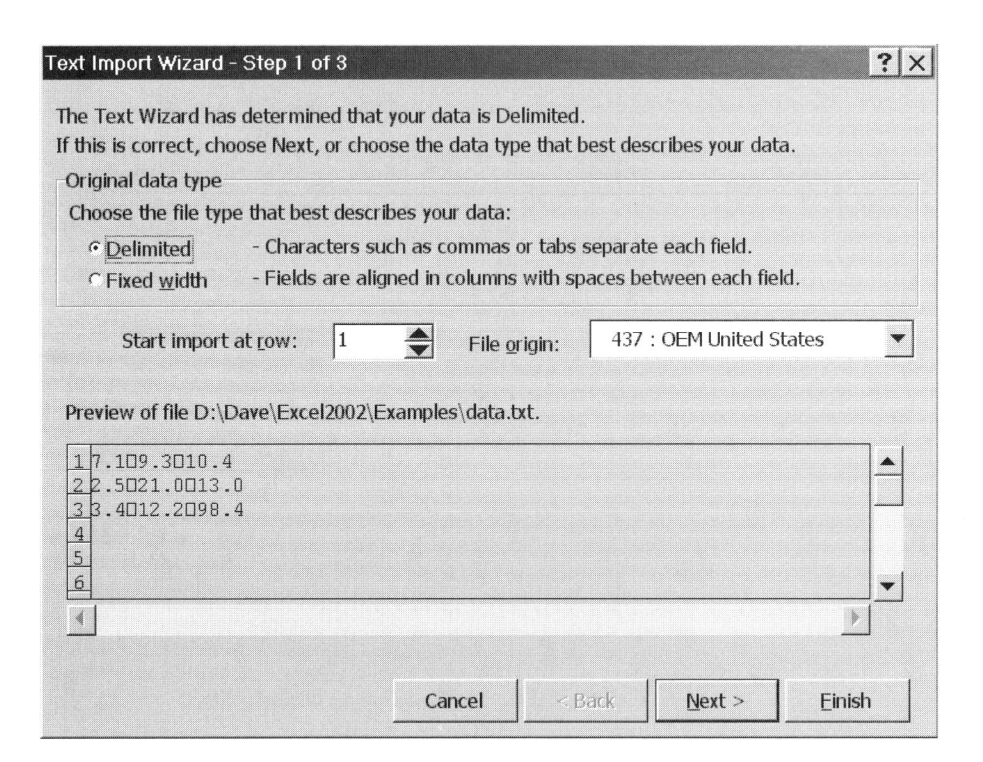

Figure 7.19. The Text Import Wizard—Step 1.

Step 7: The Text Import Wizard will show a preview of the data to be imported. The Wizard will try to guess whether the data are delimited. Note that the tab character causes the data to appear in the Preview area as being in boxes.

Step 8: Choose **Next**, and Step 2 of the Text Import Wizard will appear, as shown in Figure 7.20.

Step 9: In this dialog box you can try different delimiters. In Figure 7.20, Excel has correctly guessed that the delimiter is a tab.

Step 10: Note the box titled *Treat consecutive delimiters as one*. This feature is useful if you accidentally used more than one tab between some data elements.

Step 11: Note the box titled *Text qualifier*. In some cases, you may use a delimiter as part of the data. For example, if commas were used as delimiters, then what would happen if the text *Smith, Jr.* was entered? The comma would be treated as a delimiter instead of as part of the data. You can fix this by using the box titled *Text qualifier*. Use the text qualifier to surround data that contains a delimiter (e.g., "*Smith, Jr.*").

Step 12: Choose **Next**, and Step 3 of the Wizard will appear, as shown in Figure 7.21.

Step 13: Choose the cell format for each column of data and then choose **Finish**. An Excel worksheet will be created with the imported data.

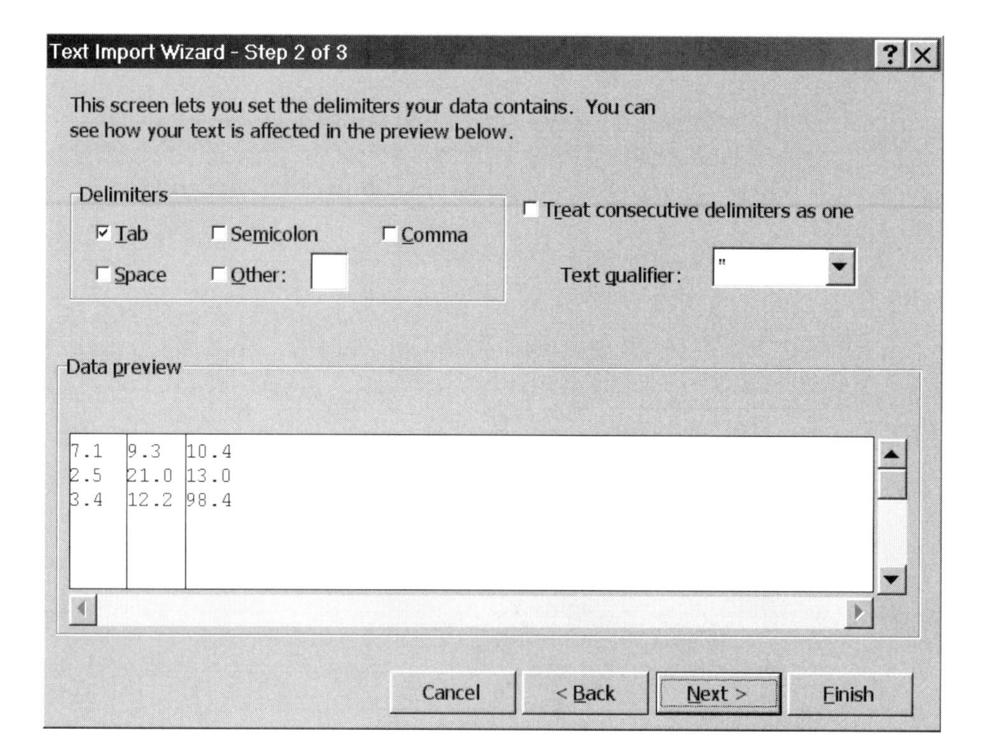

Figure 7.20. The Text Import Wizard—Step 2.

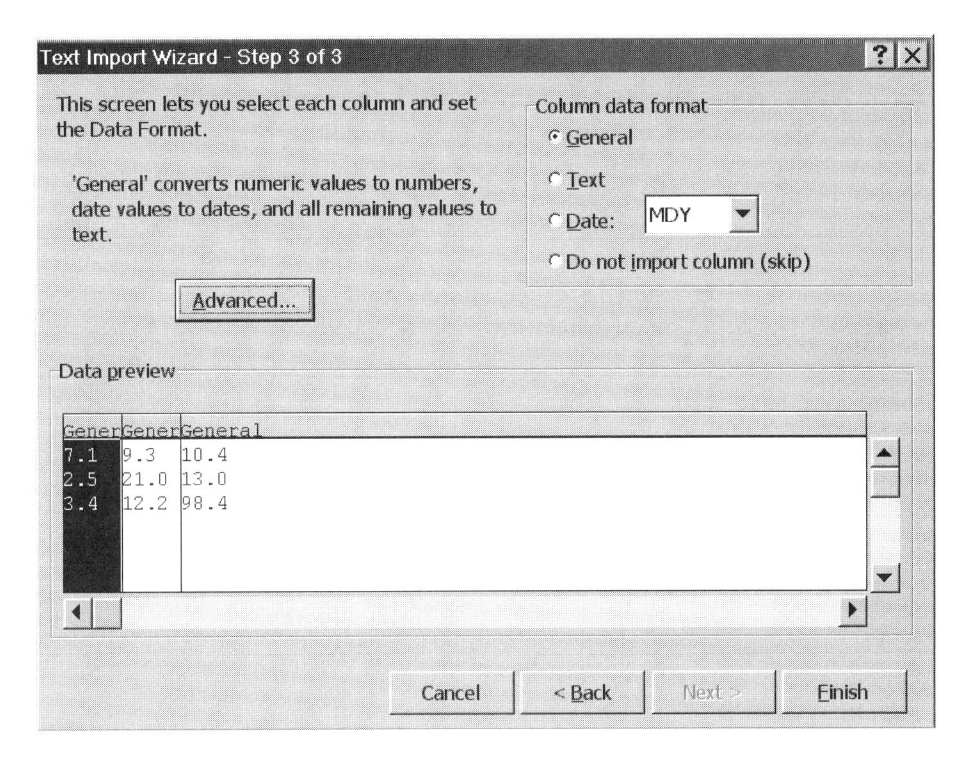

Figure 7.21. The Text Import Wizard—Step 3.

PROFESSIONAL SUCCESS—TEAM MEETING GUIDELINES

Read the following job listings:

Mechanical Engineer

We have an immediate need for an engineer to interface an engineering automation & optimization environment with a variety of CAD and CAE systems . . . B.S./M.S. in mechanical engineering . . . Good communication skills and a strong interest to *interact with customers in problem-solving situations* is a must.

Electrical Engineer

Required degree: BSEE+. Perform audio subsystem validation to verify prototypes throughout the product-development program cycle. *Must work well in a team environment.*

Aerospace Engineer

Applicants selected may be subject to a government security investigation and meet eligibility requirements for access to classified information. *Must be a team player* and possess excellent written and oral communication skills.

Industrial Engineer

Investigate manufacturing processes in continuous-improvement environment; recommend refinements. Design process equipment to improve processes. Must be highly skilled at planning/managing and be *able to sell ideas to team members* and company management.

Extrusion Engineer

This manufacturer of fiber optics is seeking an Extrusion Engineer who can handle the majority of the technical issues in production. The successful candidate must be able to *work with other disciplines in a team atmosphere* of mutual support.

Software Engineer

Looking for a well-rounded software engineer with strong experience in object-oriented design and GUI development . . . Must be a highly motivated self-starter who *works well in a team environment*.

All of the positions listed were taken from actual job postings. What do the position announcements have in common? Teamwork! The ability of an engineer to work well in the team environment has as much to do with professional success as do scientific and technical skills. Few engineering accomplishments are produced in isolation. The following guidelines may help as you begin to conduct and participate in team meetings.

Decision Making

At first, attempt to make decisions by consensus. If that fails due to a single member who continually disagrees, then move to consensus minus one approach. If consensus minus one fails, then move to a majority rule. Be aware that, as you lose the consensus of team members, the ability of the team to succeed is weakened. Time spent on gaining consensus by winning all team members over to an idea is time well spent!

Confidentiality

Respect the other members' right to confidentiality. Lay ground rules about what material, if any, is to be treated confidentially. In the world of business and government contracts, you may be asked to sign a *confidentiality agreement*. These legal documents specify which of your employer's materials are protected from disclosure.

Attention

Actively listen—ask questions or request clarification of other members' comments. Reflect a summary of important points that other team members have made. Give acknowledgement that you have understood. Try not to mentally rehearse what you are going to say while others are speaking.

Preparation

Be adequately prepared for the meeting.

Punctuality

Be on time. If you are 10 minutes late and there are six other members in the group, then you are wasting one person-hour of time!

Ensure Active Contribution

If not all team members are contributing and actively participating, then something is wrong with the group process. Stop the meeting, and take time to get everyone involved before proceeding.

Record Keeping

Someone on the team should be keeping records of team meetings.

Flexibility

One of the aspects of working on a team is that you win some and you lose some. Not every one of your ideas will be accepted by the group. Be prepared to think of creative solutions that every team member can accept.

Dynamics

Help improve relationships among the team members. Do not dominate the meeting or let another member dominate the meeting. If this cannot be resolved within the group, enlist the help of an outside *facilitator*. A facilitator is a nongroup member who does not take part in the content of the group issues. The facilitator exists to help smooth the group process. One of the roles of a facilitator is to prevent any single team member from dominating the others.

Quorum

Establish at the onset what a team quorum will be. Do not hold team meetings unless a quorum is present.

APPLICATION—USING TEAMWORK TO SOLVE A PROBLEM

Three students, Sara, Justin, and Allison, have been assigned a group project. Their assignment is to determine the energy required to pump water through a packed-bed filter for a local industrial facility. (See Figure 7.22.)

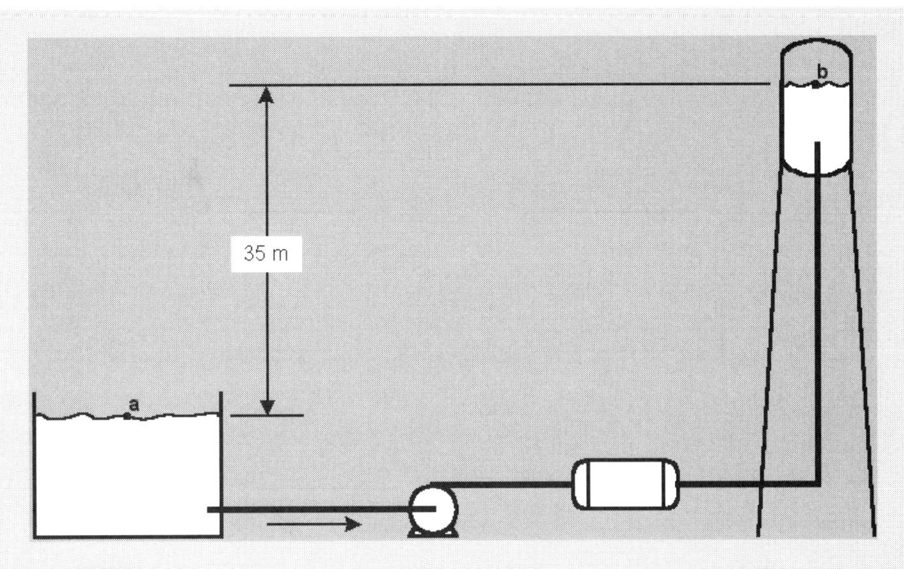

Figure 7.22. Water pumped through a packed-bed filter.

Here's what the team knows:

- Water flows at a rate of 400 liters per minute from a holding tank that is open to the atmosphere (pressure = 1 atmosphere = $1.01 \times 10^5 Pa$) and into a centrifugal pump (25 HP, 60% efficiency).
- The water then flows through the packed-bed filter and into an elevated, pressurized storage tank (pressure = $3.00 \times 10^5 Pa$).
- The elevation change between the water level in the open tank (marked a) and the water level in the pressurized tank (marked b) is 35m.
- The equation used to solve problems of this type is called the *mechanical energy balance* equation and is shown in Figure 7.23.
- The pump-energy term on the left side of the equation represents the energy added to the system by the pump (expressed in *head* (height) of fluid—civil-engineering style).
- The terms on the right side represent the energy required to change the pressure of the fluid, to lift the fluid from H_a to H_b, to accelerate the fluid from V_a to V_b, and to overcome friction (F). The group's task is to find F.

The team devises the following strategy to solve the problem:

1. The first thing they do is to throw out the acceleration term, since the fluid velocity at

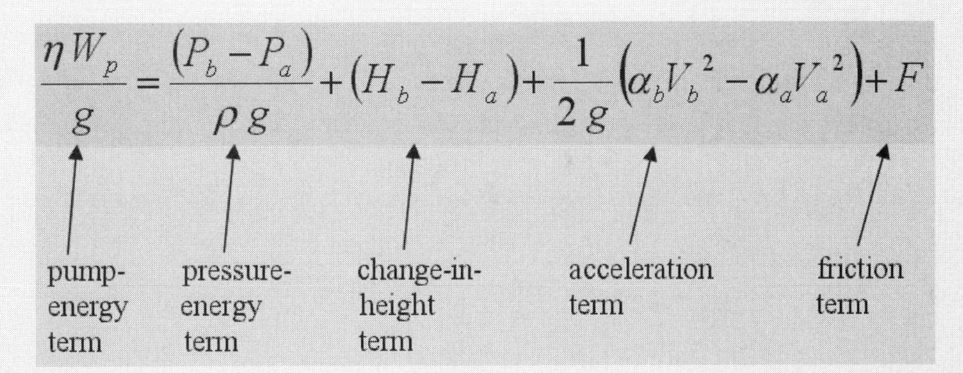

Figure 7.23. Mechanical-energy balance equation.

the surface of each tank (at points *a* and *b*) will be very small. When those small velocities are squared, the acceleration term will be insignificant.

2. That leaves the pump-energy term, the pressure-energy term, and the change-in-height term.
3. The students decide to each take one piece of the equation. They will each solve a piece of the larger problem in a separate workbook. Then they will combine their results by linking to each workbook to find *F*. Sara takes the pump term, Justin takes the pressure term, and Allison takes the elevation term.

Sara's Part

Sara's part is the most difficult. She is to find the pump-energy term, or $\eta W_p/g$.

W_p is the energy per unit mass to the pump (usually from a motor), and ηW_p is the energy per unit mass from the pump to the fluid. Sara knows that the efficiency, η, is 60% or 0.60. But she doesn't know the energy per unit mass, W_p. She also knows the power rating of the pump. Power is related to W_p through the mass flow rate through the pump:

$$\text{Power} = m_{\text{flow}} W_p.$$

Mass flow rate is related to the stated volumetric flow rate through the fluid density, ρ:

$$m_{\text{flow}} = V_{\text{flow}}\, \rho.$$

The gravitational constant $g = 9.8\ m/sec^2$ at sea level. So Sara develops her worksheet, which looks like Figure 7.24. Sara saved her worksheet with the name PumpEnergy.xls.

The equations in Sara's worksheet are as follows:

Cell B9:	=B2/0.001341	(the constant 0.001341 converts HP to Watts)
Cell B11:	=B4/1000/60	(the constant 1000 converts liters to m^3, and the constant 60 converts minutes to seconds)
Cell B14:	=B12*B11	(density times volumetric flow rate)
Cell B15:	=B9/B14	(pump power rating divided by mass flow rate)
Cell B17:	=B10*B15/B13	($\eta W_p/g$)

	A	B	C
1	**Pump Term**		
2	Power:	25	HP
3	η:	0.6	
4	V flow:	400	liters/min
5	Density(ρ):	1000	kg/m^3
6	g:	9.8	m/sec^2
7			
8	**Converting all units to SI units:**		
9	Power:	18642.8	Joules/sec
10	η:	0.6	
11	V_{flow}:	0.006667	m^3/sec
12	Density(ρ):	1000	kg/m^3
13	g:	9.8	m/sec^2
14	m_{flow}:	6.666667	kg/sec
15	W_p:	2796.421	Joules/kg
16			
17	**Pump term:**	171.2094	J·s^2 / m·kg = m

Figure 7.24. Sara's worksheet finds the pump-energy term.

Justin's Part

Justin's spreadsheet is a bit simpler, since there are fewer conversions and calculations required. He is to find the pressure-energy term = $(P_a - P_b)/\rho \cdot g$. Justin's worksheet is shown in Figure 7.25. Justin saved his worksheet with the name PressureEnergy.xls.

The equation in Justin's worksheet is

Cell B8: = (B3-B2)/(B4*B5).

Allison's Part

Allison's task is the easiest of all. She is to find the elevation term $H_b - H_a$. Allison's worksheet is shown in Figure 7.26. Allison saved her worksheet with the name Elevation.xls.

The equation in Allison's worksheet is

Cell B6: = B4-B3.

Combining the Results

Once each of the members had completed the assigned portion, the group got together and quickly finished the project. They created a summary workbook and created links to cells in their individual workbooks. Figure 7.27 shows the results.

The links that refer to other workbooks are shown next. All of the workbooks were placed in the same directory. The format for a remote cell reference is the workbook file name, followed by an exclamation point, followed by the cell number. If a reference is made to a workbook in another directory, then the link must contain the full pathname of the file location:

Cell B3: = PumpEnergy!B17;
Cell B4: = PressureEnergy!B8;
Cell B5: = Elevation!B6;
Cell B7: = B3-B4-B5.

While this is a very simple example of using a spreadsheet for collaborating on group assignments, it does illustrate how easily the results from different members can be combined to complete a group assignment.

	A	B	C
1	**Pressure Term**		
2	P_a:	1.01E+05	Pa = N/m^2
3	P_b:	3.00E+05	Pa = N/m^2
4	Density (ρ):	1000	kg/m^3
5	g:	9.8	m/s^2
6			
7	**Pressure term:**	20.30612	N·s^2 / kg = m

Figure 7.25. Justin's worksheet finds the pressure-energy term.

	A	B	C
1	**Elevation Term**		
2			
3	H_a:	0	m
4	H_b:	35	m
5			
6	**Elevation Term:**	35	m

Figure 7.26. Alison's worksheet finds the elevation term.

	A	B	C
1	**Friction Loss**		
2			
3	Pump Term:	171.2094	m
4	Pressure Term:	20.30612	m
5	Elevation Term:	35	m
6			
7	**Friction Term:**	115.9033	m

Figure 7.27. Summary worksheet finds the friction term.

KEY TERMS

change history
revision marks
comments
text qualifier
open access
write access

SUMMARY

In this chapter, you were shown the tools that Excel provides to promote team collaboration. These include methods for tracking revisions, sharing workbooks, inserting comments, and importing data from other applications. The chapter also explains the various methods of data protection that Excel provides.

Problems

1. Create a workbook with the data shown in Figure 7.2. Make two copies of the workbook. Make changes to each of the three documents. Merge the revised documents into a single document by opening the copy of the shared workbook into which you want to merge changes from another workbook file on disk. Then choose **Tools → Merge Workbooks**. Select the shared workbook to be merged and then click **OK**. Repeat these steps for both copies that are to be merged. You will be guided through the process of accepting and rejecting revisions.
2. Turn on the AutoSave feature to automatically save your document every 2 minutes. Is a new version created every 2 minutes? You can check this in the following manner:

   Step 1: Create a workbook with the data in Figure 7.2.
   Step 2: Save the workbook.
   Step 3: Open the same workbook and edit it for 5 or 6 minutes.
   Step 4: Close Excel *without* manually saving your changes. Then reopen the workbook to see if your editing was automatically saved.

3. Do the password-protection mechanisms discussed in this chapter prevent another student from making a copy of your paper? Do any of the protection methods presented in the chapter prevent someone from printing your document without knowing the password? If so, which ones?

4 Turn sharing on and create a change history for a workbook. Then turn sharing off and see if the change history is actually deleted.

5. Set the change-history timer in the Share Workbook dialog box for one day. Wait more than 24 hours and see if the history really expires.

6. Create three workbooks, one for each of Sara's, Justin's, and Allison's parts of the Pump application in this chapter. Create a fourth summary workbook that references cells in the other three workbooks and produces the final result (the friction term).

8

Excel and the World Wide Web

8.1 ENGINEERING AND THE INTERNET

The Internet is one of the primary means of communication for scientists and engineers. Correspondence through electronic mail, the transfer of data and software via electronic file transfer, and research by using on-line search engines and databases are everyday tasks for engineers. The World Wide Web (WWW or simply Web) is a collection of technologies for publishing, sending, and obtaining information using the Internet. The Internet requires every engineering student to learn two new essential skills. One, every student must gain fluency in searching, locating, and retrieving relevant technical information from the WWW. Second, every engineering student must learn how to post written documents to the WWW. The ability to present technical results via the WWW is an essential communication skill for modern engineers. In this chapter, we will focus on the first task—the retrieval of documents in relation to Excel.

SECTION

OBJECTIVES

After reading this chapter, you should be able to:

- Access the World Wide Web from within an Excel worksheet
- Obtain extra templates and other add-ins from the WWW
- Retrieve files from HTTP servers into a local worksheet
- Use the Web Query feature to import Excel data from the WWW
- Create hyperlinks in a worksheet
- Convert Excel documents to HTML

PROFESSIONAL SUCCESS

The World Wide Web holds a wealth of information about your new profession. Take some time to visit the professional societies that represent your discipline. The following URLs represent a few of the national and international organizations that are on-line.

Accreditation Board for Engineering and Technology (ABET)	**http://www.abet.org/ABET.html**
American Institute for Aeronautics and Astronautics (AIAA)	**http://www.aiaa.org**
American Institute of Chemical Engineers (AICHE)	**http://www.aiche.org**
American Society of Civil Engineers (ASCE)	**http://www.asce.org**
American Society of Engineering Education (ASEE)	**http://www.asee.org/asee**
American Society of Mechanical Engineers (ASME)	**http://www.asme.org**
American Society of Naval Engineers (ASNE)	**http://www.jhuapl.edu/ASNE**
Institute of Electrical and Electronics Engineering (IEEE)	**http://www.ieee.org**
National Society of Black Engineers (NSBE)	**http://www.nsbe.org**
National Society of Professional Engineers (NSPE)	**http://www.nspe.org**
Society of Women Engineers (SWE)	**http://www.swe.org**

8.2 ACCESSING THE WORLD WIDE WEB FROM WITHIN EXCEL

To access the Internet from within Excel, your computer must be connected to the Internet. If you are in a computer lab at school, then the computer may be connected to a local area network (LAN) through a network card. The LAN may or may not be connected to the Internet. Ask your lab manager or instructor for details.

If your computer is not directly connected to a LAN, then you can access the Internet via a modem. This is called *dial-up networking*. For information about dial-up networking, access the Help section on the Task bar. Locate the topic titled *Dial-Up Networking*. The computer systems group at your institution may be able to help you with some of the details, such as the assignment of an IP address, netmasks, etc. During the rest of this chapter, it is assumed that your computer is connected to the Internet.

Turn on the Web toolbar by choosing **View → Toolbars** from the Menu bar. Check the item titled *Web*. The Web toolbar is shown in Figure 8.1.

Figure 8.1. The Web toolbar.

To open a Web page or local Web document,

Step 1: Choose the **Go** icon Go from the Web toolbar. A drop-down menu will appear.
Step 2: Select **Open Hyperlink** from the drop-down menu.
Step 3: The Open Internet Address dialog box will appear, as shown in Figure 8.2.

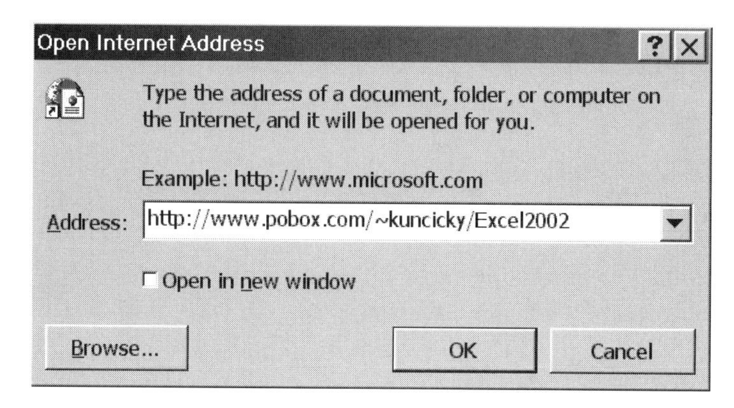

Figure 8.2. The Open Internet Address dialog box.

From the Open Internet Address dialog box, you can type or choose a remote or local Web page. The address of a remote Web page is called a *Uniform Resource Locator* or *URL*. Here's an example of a URL:

http://www.pobox.com/~kuncicky/Excel2002.

The meanings of the components of this URL are:

http	This stands for *HyperText Transfer Protocol*.
www.pobox.com	This is the name of a Web server.
~kuncicky	This refers to the space allocated to a user named Kuncicky.
Excel2002	This is the path of a Web page or folder.

Figure 8.2 shows the URL for the author's Web pages for this book. You can also type in the path and name of a local Web document, or select the local document by choosing the **Browse** button.

Previously accessed sites can be viewed by using the small arrow at the right side of the Web toolbar. This is shown in Figure 8.3.

A Web document is written in a markup language called *Hypertext Markup Language* or *HTML*. The HTML document is usually viewed via an application called a Web browser. The Web browser interprets the HTML document and displays the results as text, graphics, animations, sounds, etc.

Excel acts as a front end to a Web browser. When you click **OK**, the selected page will be displayed through your default Web browser, such as Microsoft's Internet Explorer or Netscape. Since you may already be familiar with your favorite browser, you may want to use that browser's features directly, instead of the Excel front end, for general Web browsing.

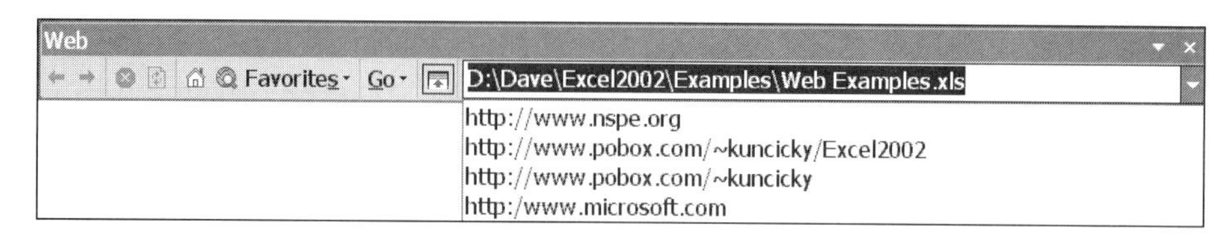

Figure 8.3. Viewing previously accessed sites.

8.3 WEB SITES RELATED TO MICROSOFT EXCEL

Microsoft maintains a section of Web pages specifically for Excel users. The primary page for Microsoft is located at

http://www.microsoft.com.

Within the Microsoft site are an Excel tutorial, product information, and a number of free add-ins, patches, and templates. It is important to periodically check for updates to your programs, since security problems are frequently found in most popular applications.

8.4 RETRIEVING DATA BY USING A WEB QUERY

The Web Query feature retrieves data from an external source over the Web and places the data in a local Excel worksheet. To run a Web Query,

Step 1: Open a new blank workbook.

Step 2: Choose **Data → Import External Data → Import Data** from the Menu bar. The Select Data Source dialog box will appear, as shown in Figure 8.4.

Step 3: Several sample queries are provided with the standard Excel installation. If you would like to download more Web Queries, choose **Get More Web Queries**.

Step 4: As an example, select **MSN MoneyCentral Investor Currency Rates**.

Step 5: Choose the **Open** button. The Import Data dialog box will appear, as shown in Figure 8.5.

Step 6: Choose a location for the query results. Then press **OK**. The current rates will be placed in your worksheet.

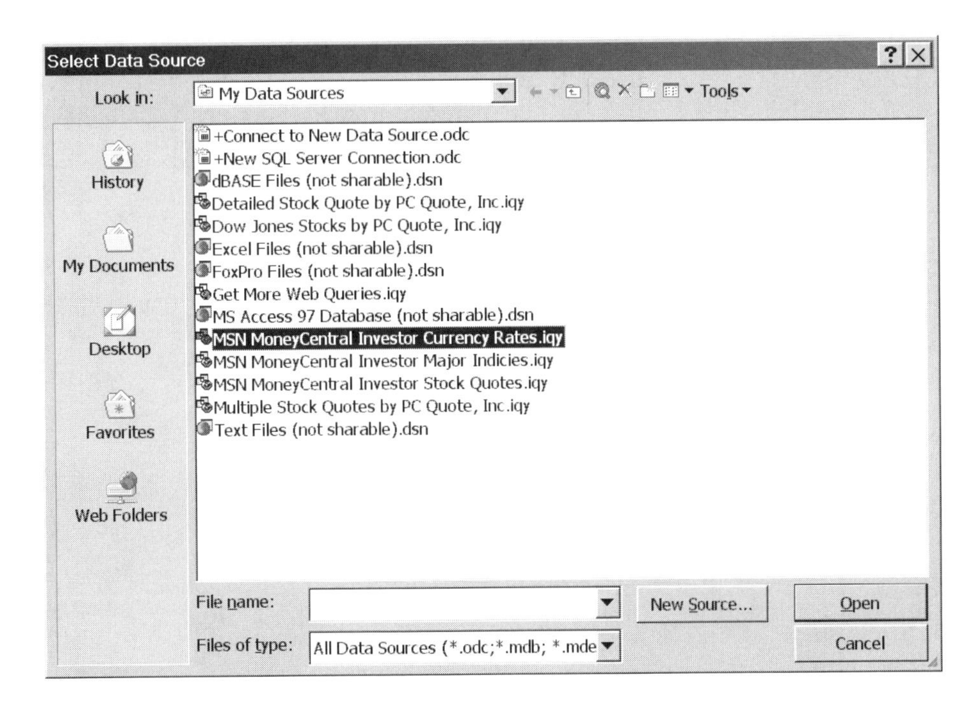

Figure 8.4. The Select Data Source dialog box.

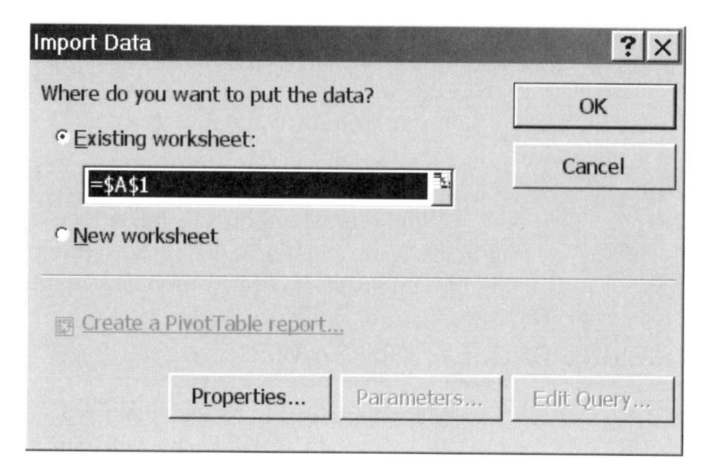

Figure 8.5. The Import Data dialog box.

A *Web Query* is a formatted text file. The contents of the MSN query used in the previous example are displayed in Figure 8.6. The effect of executing the query is to access the Web server at the URL **http://moneycentral.msn.com** and execute the command named *rates.asp*. The results are returned and displayed in your local Excel worksheet.

Choose the **Properties** button of the Import Data dialog box. The External Data Range Properties dialog box will appear, as shown in Figure 8.7.

From this dialog box, you can modify a number of query options, such as the ability to automatically refresh or update the data. In Figure 8.7, the box titled *Enable background refresh* is checked and the refresh rate is set to 60 minutes. As long as you have your worksheet open and you are connected to the Internet, the Web Query will automatically run every 60 minutes and update the currency rates.

Part of the results from the currency query are displayed in Figure 8.8.

```
WEB
1
http://moneycentral.msn.com/investor/external/excel/rates.asp

Selection=EntirePage
Formatting=All
PreFormattedTextToColumns=True
ConsecutiveDelimitersAsOne=True
SingleBlockTextImport=False
```

Figure 8.6. Contents of MSN Currency Rates Web Query.

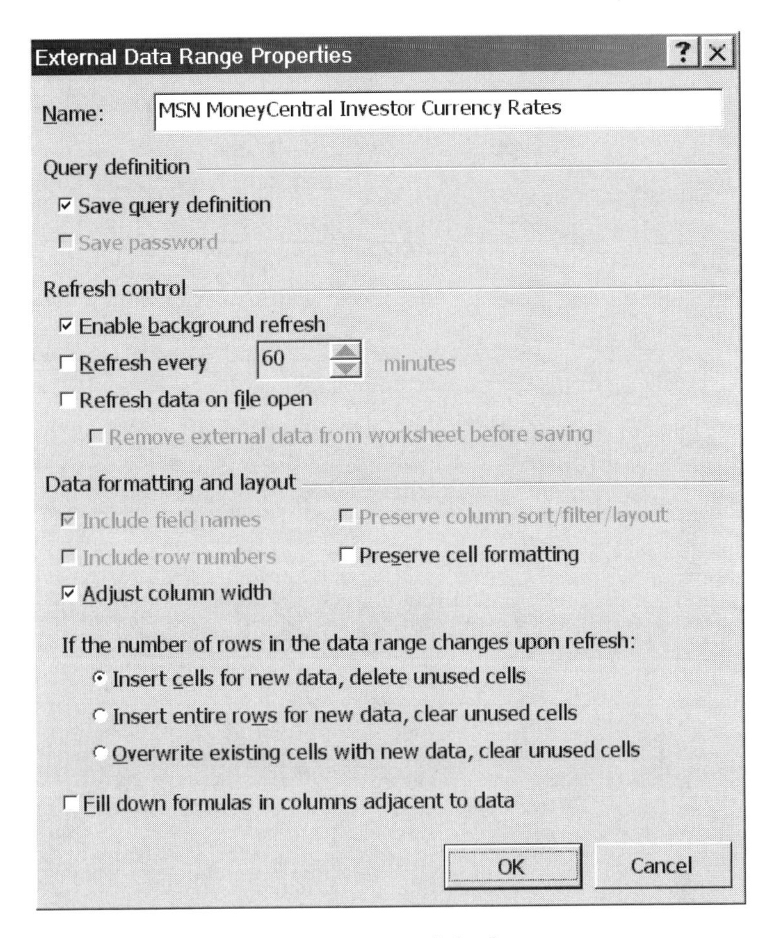

Figure 8.7. The External Data Range Properties dialog box.

	A	B	C
1	Currency Rates Provided by MSN Money		
2	Click here to visit MSN MoneyCentral Investor		
3			
4	Name	In US$	Per US$
5	Argentine Peso	0.45045	2.22
6	Australian Dollar	0.5211	1.919
7	Austrian Schilling	0.06373	15.691
8	Belgian Franc	0.02174	45.99
9	Baharain Dinar	2.6526	0.377
10	Bolivia Bolivano	0.143	6.993
11	Brazilian Real	0.42553	2.35

Figure 8.8. Results from the MSN Currency Rates Web Query.

8.5 ACCESSING EXCEL FILES ON THE WEB

To download a workbook from a remote Web site,

Step 1: Open your Web browser. We will use Internet Explorer as a sample browser. (The commands for Netscape are similar.)

Step 2: In the box titled address, type a URL. For example, to access the author's download site, type

http://www.pobox.com/~kuncicky/Excel2002.

Step 3: Select a worksheet. For example, choose the **Pump Energy Workbook** from the author's site.

Step 4: If you choose the **Open** button, Excel will start and open the Pump Energy workbook.

Step 5: If you choose the Save button, you will be prompted to enter a local file location in which to save the workbook.

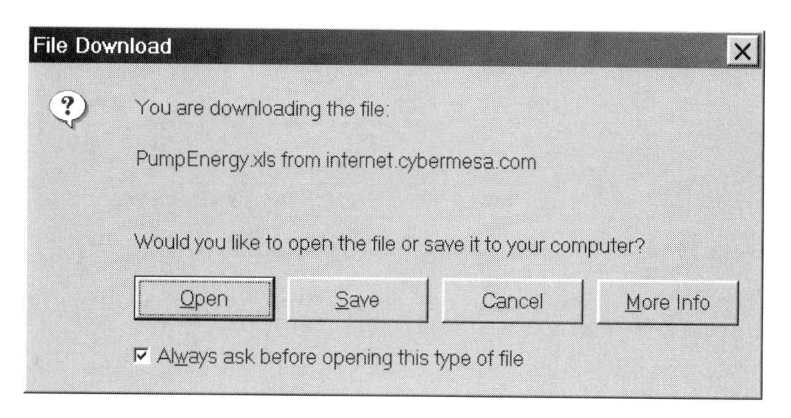

Figure 8.9. The File Download dialog box.

8.6 CREATING HYPERLINKS WITHIN A WORKSHEET

A *hyperlink*, or simply *link*, can be thought of as a pointer to another document. When you click on a hyperlink, that document is immediately displayed. The linked document may be another Excel worksheet on your local computer, or it may be a document from another application, such as Microsoft Word. If the linked document belongs to another application, then that application is started automatically for you.

A link may also point to a remote document that is retrieved from the World Wide Web, using the HTTP protocol. As more computers are connected to the Internet and as network speeds increase, the difference in the time necessary to access a local document and a remote document will diminish.

The method for creating hyperlinks is the same for local or remote documents. The only difference is the address or path name of the document. The instructions for how to create a local link and a remote link will follow.

To create a local link in a worksheet,

Step 1: Create two workbooks that resemble Figures 8.10 and 8.11.

Step 2: Save the workbook that resembles Figure 8.10 with the name **SI Units.xls**.

Step 3: Save the workbook that resembles Figure 8.11 with the name **SI Base Units.xls**.

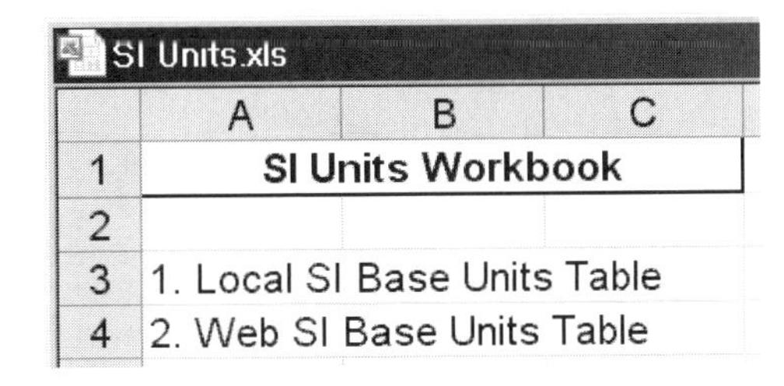

SI Units.xls

	A	B	C
1	SI Units Workbook		
2			
3	1. Local SI Base Units Table		
4	2. Web SI Base Units Table		

Figure 8.10. Main SI Units Workbook.

Microsoft Excel - SI Base Units.xls

	A	B	C
1	SI Base Units		
2	Quantity	Name	Symbol
3	length	meter	m
4	mass	kilogram	kg
5	time	second	s
6	electric current	ampere	A
7	temperature	kelvin	K
8	amount of substance	mole	mol
9	luminous intensity	candela	cd

Figure 8.11. SI Base Units.

Step 4: Open the workbook **SI Units.xls**.

Step 5: Select cell A3 (titled *Local SI Base Units Table*) and right click. A drop-down menu will appear.

Step 6: Choose **Hyperlink** from the drop-down menu. The Insert Hyperlink dialog box will appear, as shown in Figure 8.12.

Step 7: Select the workbook named **SI Base Units.xls** from the list. You may have to browse the file system to locate the file.

Step 8: Click **OK**. You will be returned to your worksheet.

The contents of cell A3 will have changed color and will be underlined, indicating that this cell contains a hyperlink. Click the hyperlink, and the referenced file will appear. You can toggle back and forth between the source and destination files by using the left and right arrows on the Web toolbar.

To add a Web link to your workbook,

Step 1: Open the workbook named **SI Units.xls**.

Step 2: Select cell A4 (titled *Web SI Base Units Table*) and right click. A drop-down menu will appear.

Step 3: Choose **Hyperlink** from the drop-down menu. The Insert Hyperlink dialog box will appear, as shown in Figure 8.12.

Step 4: Type the following URL in the box titled *Address*:

http://physics.nist.gov/cuu/Units/units.html.

This will link cell A4 to the U.S. Government's National Institute of Standards and Technology.

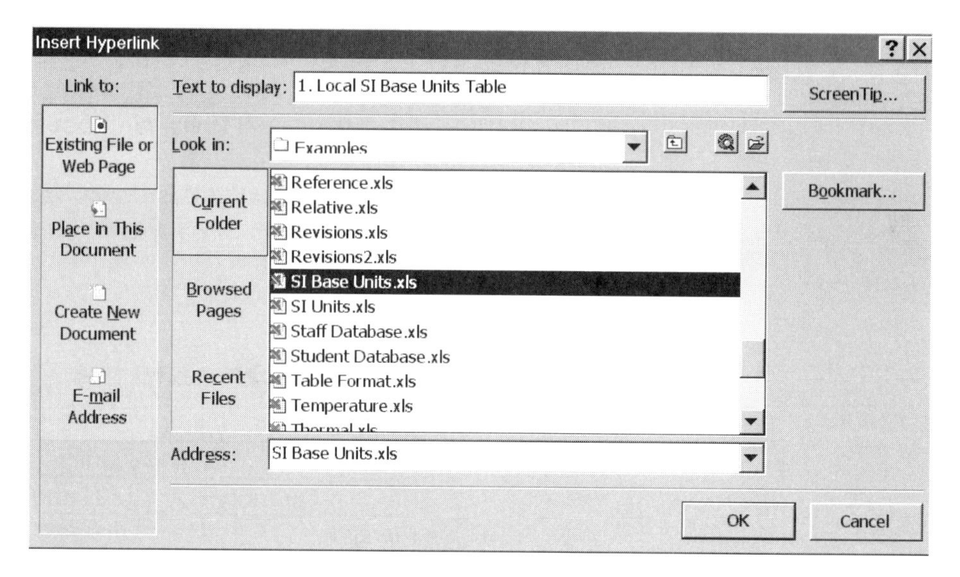

Figure 8.12. The Insert Hyperlink dialog box.

The contents of cell A4 will have changed color and will be underlined, indicating that this cell contains a hyperlink. Click the hyperlink, and the referenced Web site will be accessed via your default browser. You can toggle back and forth between the source and destination files by using the left and right arrows on the Web toolbar.

8.7 CONVERTING A WORKSHEET TO A WEB PAGE

You can post an Excel workbook on a Web server. The workbook can be downloaded or opened by using Excel. There are times when you may want to post the contents of a workbook in HTML format. This will allow users to view your Web page directly with their browser. That is, they will be able to view your Web page even if they do not have the Excel program.

To covert a worksheet to HTML,

Step 1: Open the worksheet that you want to convert. For example, open the workbook named **SI Base Units.xls**.

Step 2: Choose **File → Web Page Preview** from the Menu bar. You will be shown a preview of the Excel-to-HTML conversion.

Step 3: Excel will create a new file with the same name as your workbook, but with the extension *.htm*.

Step 4: If you are satisfied with the preview, choose **File → Save as Web Page** from the Excel Menu bar. The Save As dialog box will appear, as shown in Figure 8.13.

Step 5: From the Save As dialog box, you can select all or part of the workbook to convert.

Step 6: Note the item labeled *Add interactivity*. If you check this item, the user will be able to interact with the resulting Web page. If you do not check this item, the resulting Web page will be static. Of course, the ability to modify a page on a remote server also depends on the permissions granted to users by the server.

Once you have created an HTML version of your worksheet, you may edit the HTML document by using any HTML editor. You can open and edit the HTML file from Excel by choosing **File → Open**. Open the desired file. Note that an HTML file will have an *.htm* extension, not an *.xls* extension.

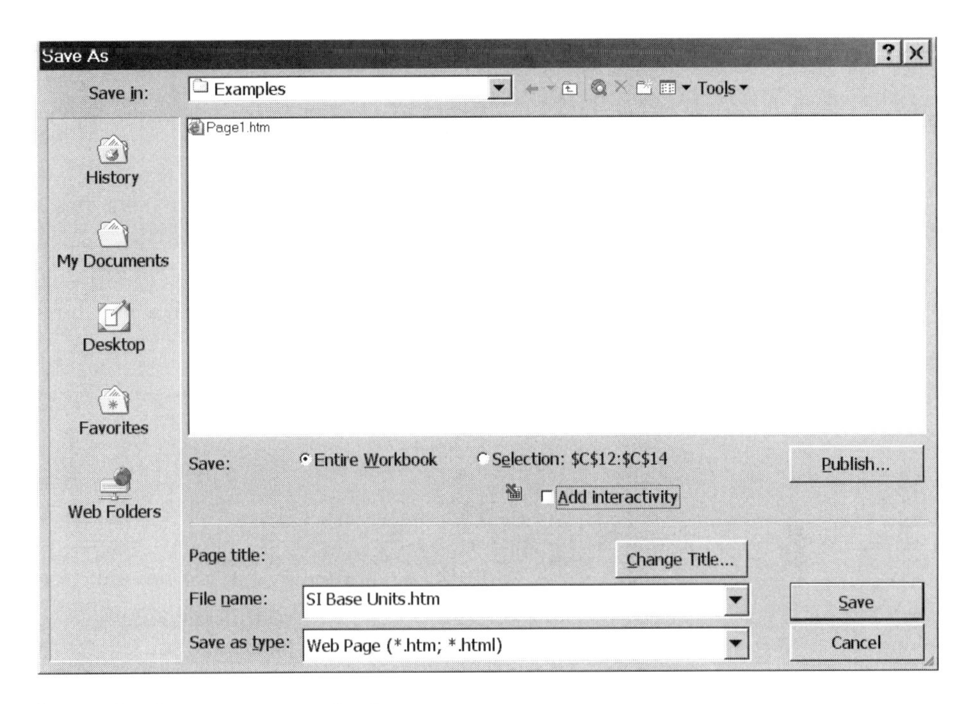

Figure 8.13. The Save As dialog box.

KEY TERMS

dial-up networking
HyperText Transfer Protocol
link
Web Query
HTML
hyperlink
Uniform Resource Locator
HTTP
HyperText Markup Language
URL

SUMMARY

This chapter introduced you to the ways that the World Wide Web and Excel can interface. Excel files from remote HTTP sites can be opened by Excel and saved to your local computer. The Web Query feature automates the retrieval of remote data. Local and Web-based hyperlinks can be added to a worksheet. Excel can be used to preview and convert worksheets to HTML documents.

Problems

1. Open the workbook that you created in this chapter named **SI Units.xls**. Add a new cell entitled *SI Base Unit Definitions*. Create a hyperlink from this cell to the following URL:

   http://physics.nist.gov/cuu/Units/current.html.

2. Use the Help feature to read about the HYPERLINK function. Create a valid hyperlink in a worksheet by using the HYPERLINK function.

3. One advantage of using the HYPERLINK function is that the link can depend on a conditional expression. Create an IF expression that links to **www.netscape.com** if cell A1 = *Netscape* and links to **www.microsoft.com** if cell A1 = *Explorer*.

Appendix A

Commonly Used Functions

ABS(*n*)	Returns the absolute value of a number
AND(*a*, *b*, . . .)	Returns the logical AND of the arguments (TRUE if all arguments are TRUE, otherwise FALSE)
ASIN(*n*)	Returns the arcsine of *n* in radians
AVEDEV(*n1*, *n2*, . . .)	Returns the average of the absolute deviations of the arguments from their mean
AVERAGE(*n1*, *n2*, . . .)	Returns the arithmetic mean of its arguments
BIN2DEC(*n*)	Converts a binary number to decimal
BIN2HEX(*n*)	Converts a binary number to hexadecimal
BIN2OCT(*n*)	Converts a binary number to octal
CALL(. . .)	Calls a procedure in a DLL or code resource
CEILING(*n*, *sig*)	Rounds a number *n* up to the nearest integer (or nearest multiple of significance, *sig*)
CHAR(*n*)	Returns the character represented by the number *n* in the computer's character set
CHIDIST(*x*, *df*)	Returns the one-tailed probability of the chi-squared distribution, using *df* degrees of freedom
CLEAN(*text*)	Removes all nonprintable characters from *text*
COLUMN(*ref*)	Returns the column number of a reference
COLUMNS(*ref*)	Returns the number of columns in a reference
COMBIN(*n*, *r*)	Returns the number of combinations of *n* items, choosing *r* items
COMPLEX(*real*, *imag*, *suffix*)	Converts real and imaginary coefficients into a complex number
CONCATENATE(*str1*, *str2*, . . .)	Concatenates the string arguments
CORREL(**A1, A2**)	Returns the correlation coefficients between two data sets
COS(*n*)	Returns the cosine of an angle
COUNTBLANK(*range*)	Counts the number of empty cells in a specified range
DEC2BIN(*n*, *p*)	Converts the decimal number *n* to binary, using *p* places (or characters)
DELTA(*n1*, *n2*)	Tests whether two numbers are equal
ISERROR(*v*)	Returns TRUE if value *v* is an error
ISNUMBER(*v*)	Returns TRUE if value *v* is a number

FACT(*n*)	Returns the factorial of *n*
FORECAST(*x, known x's, known y's*)	Predicts a future value based on a linear trend
LN(*n*)	Returns the natural logarithm of *n*
MDETERM(**A**)	Returns the matrix determinant of array **A**
MEDIAN(*n1, n2,* . . .)	Returns the median of its arguments
MOD(*n, d*)	Returns the remainder after *n* is divided by *d*
OR(*a, b,* . . .)	Returns the logical OR of its arguments (TRUE if any argument is TRUE, FALSE if all arguments are FALSE)
PI()	Returns the value of *pi* to 15 digits of accuracy
POWER(*n, p*)	Returns the value of *n* raised to the power of *p*
PRODUCT(*n1, n2,* . . .)	Returns the product of its arguments
QUOTIENT(*n, d*)	Returns the integer portion of *n* divided by *d*
RADIANS(*d*)	Converts degrees to radians
RAND()	Returns an evenly distributed pseudorandom number > = 0 and < 1
ROUND(*n, d*)	Rounds *n* to *d* digits
ROW(*ref*)	Returns the row number of a reference
SIGN(*n*)	Returns the sign of a number *n*
SQRT(*n*)	Returns the square root of a number *n*
STDEVP(*n1, n2,* . . .)	Calculates the standard deviation of its arguments
SUM(*n1, n2,* . . .)	Returns the sum of its arguments
SUMSQ(*n1, n2,* . . .)	Returns the sum of the squares of its arguments
TAN(*n*)	Returns the tangent of an angle
TRANSPOSE(A)	Returns the transpose of an array
TREND(*known y's, known x's, new x's, constant*)	Returns values along a linear trend by fitting a straight line, using the least squares method
VARP(*n1, n2,* . . .)	Calculates the variance of its arguments

Index

N

O

P

Q

R

S

T

U

V

W

X